Detector Circuits

Detector Circuits

Rudolf F. Graf

Newnes
Boston Oxford Johannesburg Melbourne New Delhi Singapore

Newnes is an imprint of Butterworth-Heinemann

ℛ A member of the Reed Elsevier group

Library of Congress Cataloging-in-Publication Data
Graf, Rudolf F.
    [Modern detector circuit encyclopedia]
    Detector circuits / Rudolf F. Graf.
      p. cm.
    Originally published: The modern detector circuit encyclopedia.
  Blue Ridge Summit, PA : TAB Books, c1993
    Includes index.
    ISBN 0-7506-9879-9
    1. Detectors—Design and construction. 2. Electronic circuits.
  I. Title.
  TK7870.G656   1996
  681'.2—dc20
                                    96-36497
                                    CIP

British Library Cataloguing-in-Publication Data
A catalogue record for this book is available from the British Library.

The publisher offers special discounts on bulk orders of this book.
For information, please contact:
Manager of Special Sales
Butterworth–Heinemann
313 Washington Street
Newton, MA 02158–1626
Tel: 617-928-2500
Fax: 617-928-2620

For information on all Newnes electronics publications available, contact
our World Wide Web home page at: http://www.bh.com/bh

Printed in the United States of America
10 9 8 7 6 5 4 3 2 1

Contents

Introduction

Like the other volumes in this series, this book contains a wealth of ready-to-use circuits that serve the needs of the engineer, technician, student and, of course, the browser. These unique books contain more practical, ready-to-use circuits focused on a specific field of interest, than can be found anywhere in a single volume.

1

Air-Flow Detectors

The sources of the following circuits are contained in the Sources section, which begins on page 214. The figure number in the box of each circuit correlates to the source entry in the Sources section.

Air-Motion Detector
Air-Flow Detector

AIR-MOTION DETECTOR

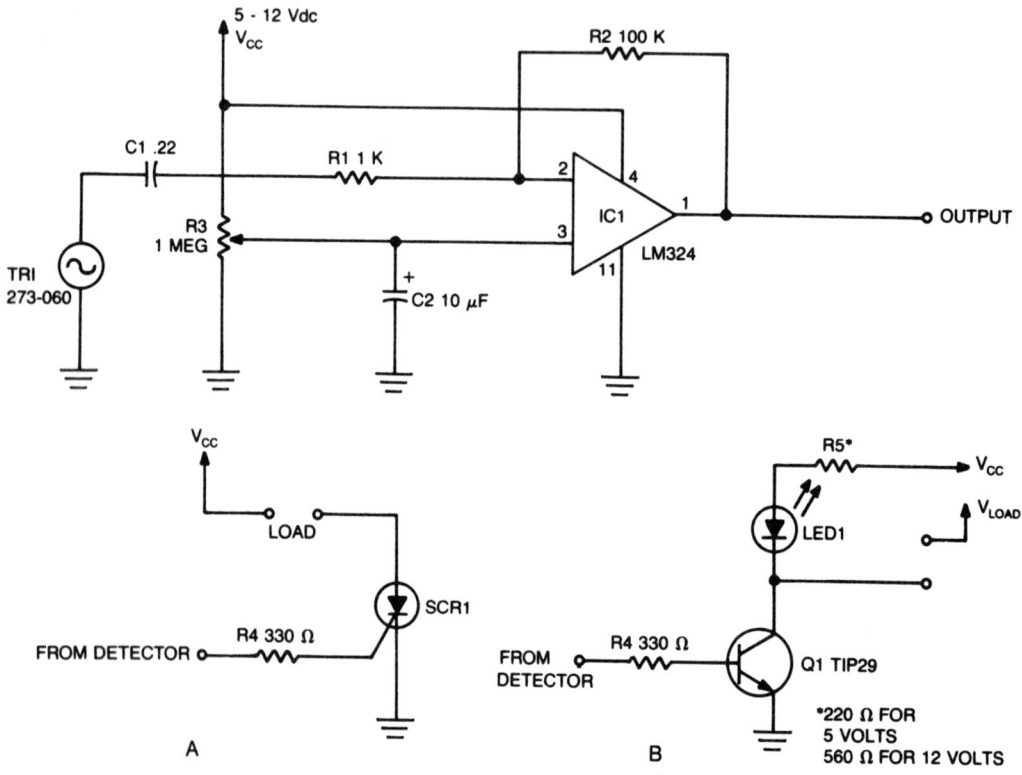

Fig. 1-1

Sensing circuit detects either steady or fluctuating air flows. The heart of the circuit is a Radio Shack piezo buzzer (P/N 273-060) and an LM324 quad op amp (The red wire from the piezo element connects to capacitor C1, and the black wire connects to ground). When a current of air hits the piezo element, a small signal is generated and is fed through C1 and R1 to the inverting input (pin 2) of one section of the LM324. That causes the output (pin 1) to go high. Resistor R3 adjusts sensitivity. The circuit can be made sensitive enough to detect the wave of a hand or the sensitivity can be set so low that blowing hard on the element will produce no output. Resistor R2 is used to adjust the level of the output voltage at pin 1. The detector circuit can be used in various control applications. For example, an SCR can be used to control 117-Vac loads, as shown in A. Also, an npn transistor, such as a TIP29, can be used to control loads, as shown in B.

AIR-FLOW DETECTOR

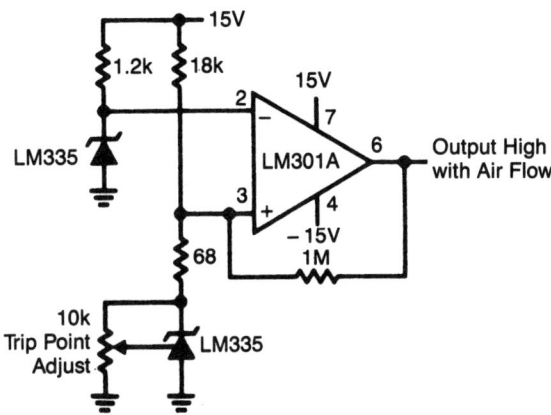

*Self heating is used to detect air flow

NATIONAL SEMICONDUCTOR

Fig. 1-2

2

Bug Detector

The source of the following circuit is contained in the Sources section, which begins on page 214. The figure number contained in the box of the circuit correlates to the source entry in the Sources section.

Bug Detector

BUG DETECTOR

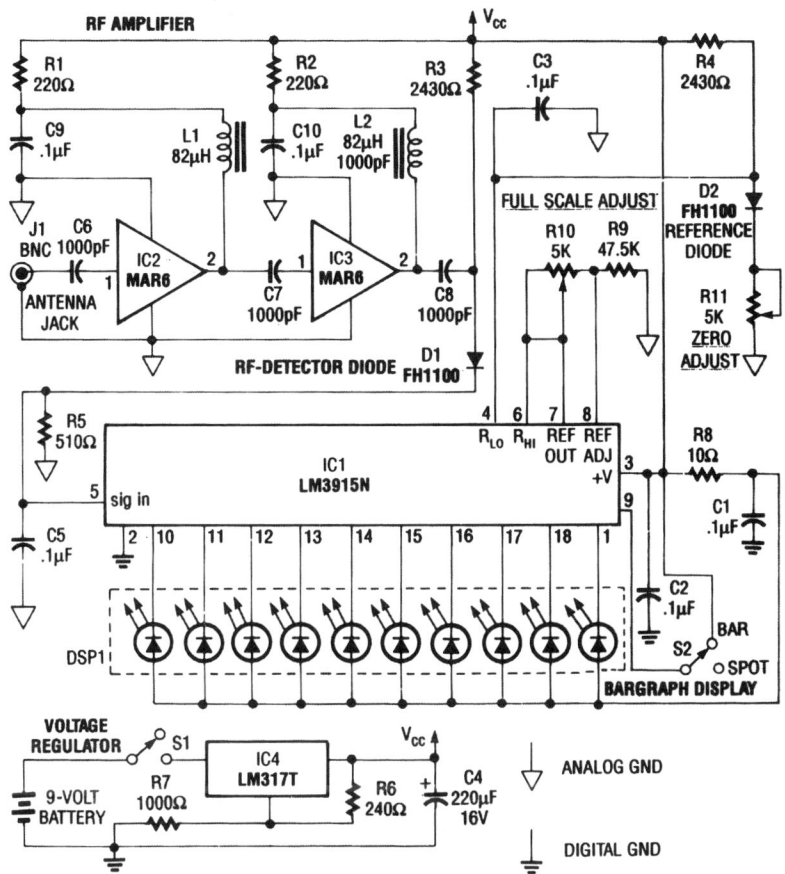

Reprinted with permission from Radio-Electronics Magazine, June 1989. Copyright Gernsback Publications, Inc., 1989.

RADIO-ELECTRONICS

Fig. 2-1

This RF detector can locate low-power transmitters (bugs) that are hidden from sight. It can sense the presence of a 1-mW transmitter at 20 feet, which is sensitive enough to detect the tiniest bug. As you bring the RF detector closer to the bug, more and more segments of its LED bar-graph display light, which aids in direction finding.

The front end has a two-stage wideband RF amplifier, and a forward-biased hot-carrier diode for a detector. After detection, the signal is filtered and fed to IC1, an LM3915N bar-graph driver having a logarithmic output. Each successive LED segment represents a 3-dB step.

3

Decoders

The sources of the following circuits are contained in the Sources section, which begins on page 214. The figure number in the box of each circuit correlates to the source entry in the Sources section.

TONE-ALERT DECODER

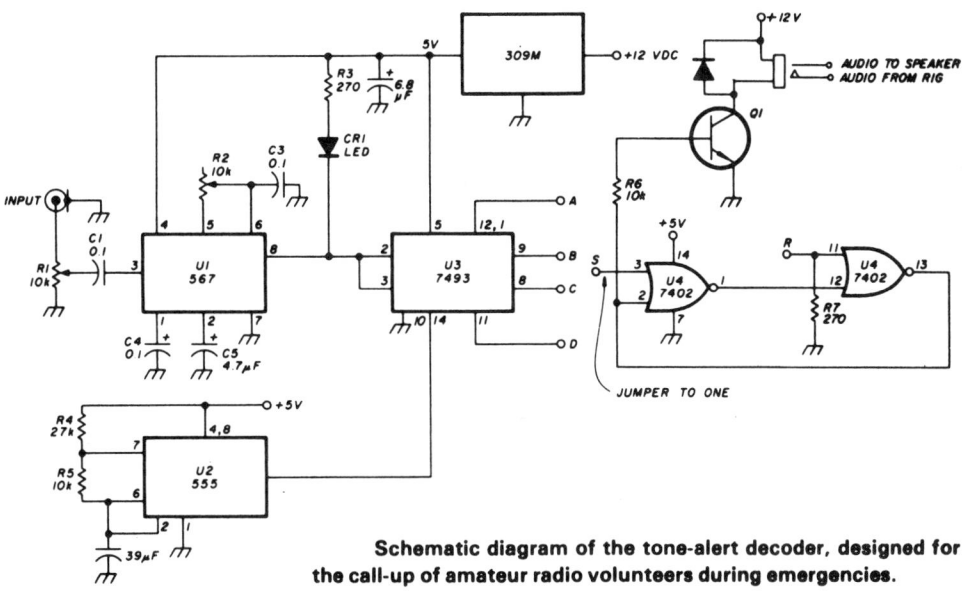

Schematic diagram of the tone-alert decoder, designed for the call-up of amateur radio volunteers during emergencies.

Fig. 3-1

PLL (U1) is set with R2 to desired tone frequency. LED lights to indicate lock-up of PLL. Reduce signal level (R1) and readjust R2 to assure lock-up. Delay is selected from counter U3 output. The circuit latches (turns on Q1 to allow audio to speaker) when a proper frequency/duration signal is received. To reset the latch, a positive voltage must be applied briefly to the *R* input of U4.

TONE DECODER WITH RELAY OUTPUT

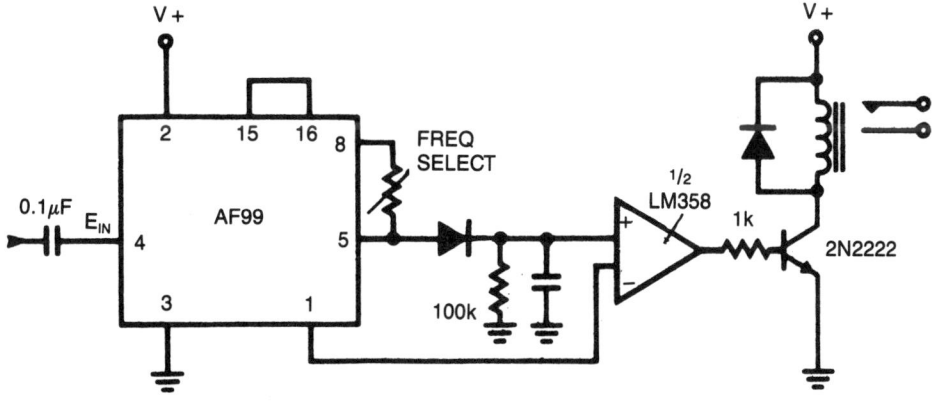

Fig. 3-2

24%-BANDWIDTH TONE DECODER

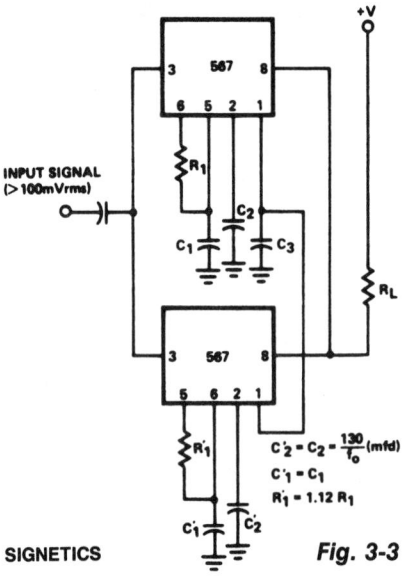

+V
567
3
8
6 5 2 1
INPUT SIGNAL
(>100mVrms)
R1
C2
C1
C3
RL
567
3
8
5 6 2 1
R'1
C'2 = C2 = 130/fo (mfd)
C'1 = C1
R'1 = 1.12 R1
C'1
C'2

SIGNETICS

Fig. 3-3

DUAL-TONE DECODER

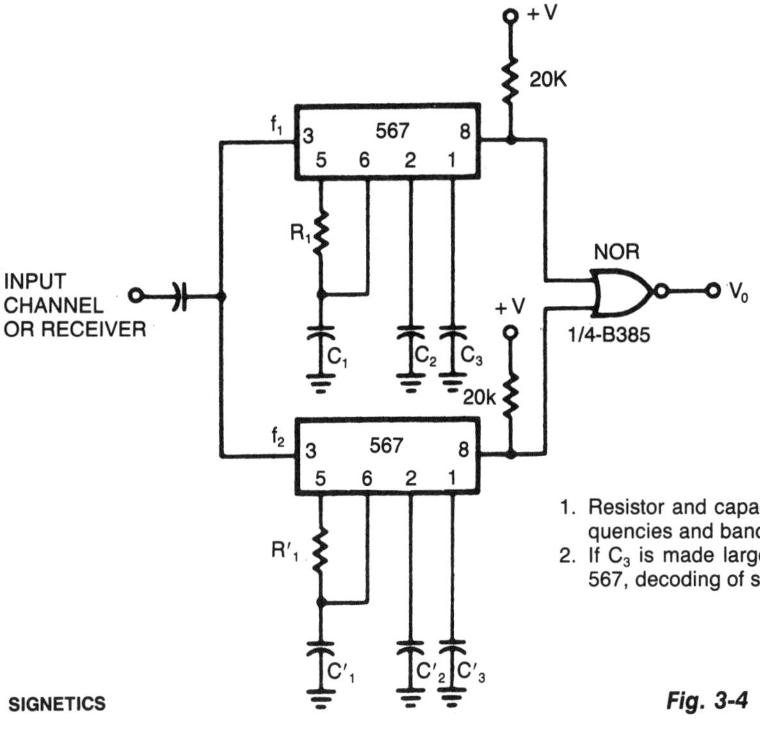

+V
20K
f1
3
567
8
5 6 2 1
R1
NOR
INPUT
CHANNEL
OR RECEIVER
+V
Vo
1/4-B385
C1
C2
C3
20k
f2
3
567
8
5 6 2 1
R'1
C'1
C'2
C'3

1. Resistor and capacitor values chosen for desired frequencies and bandwidth.
2. If C_3 is made large so as to delay turn-on of the top 567, decoding of sequential (f_1 f_2) tones is possible.

SIGNETICS

Fig. 3-4

TIME-DIVISION MULTIPLEX (TDM) STEREO DECODER

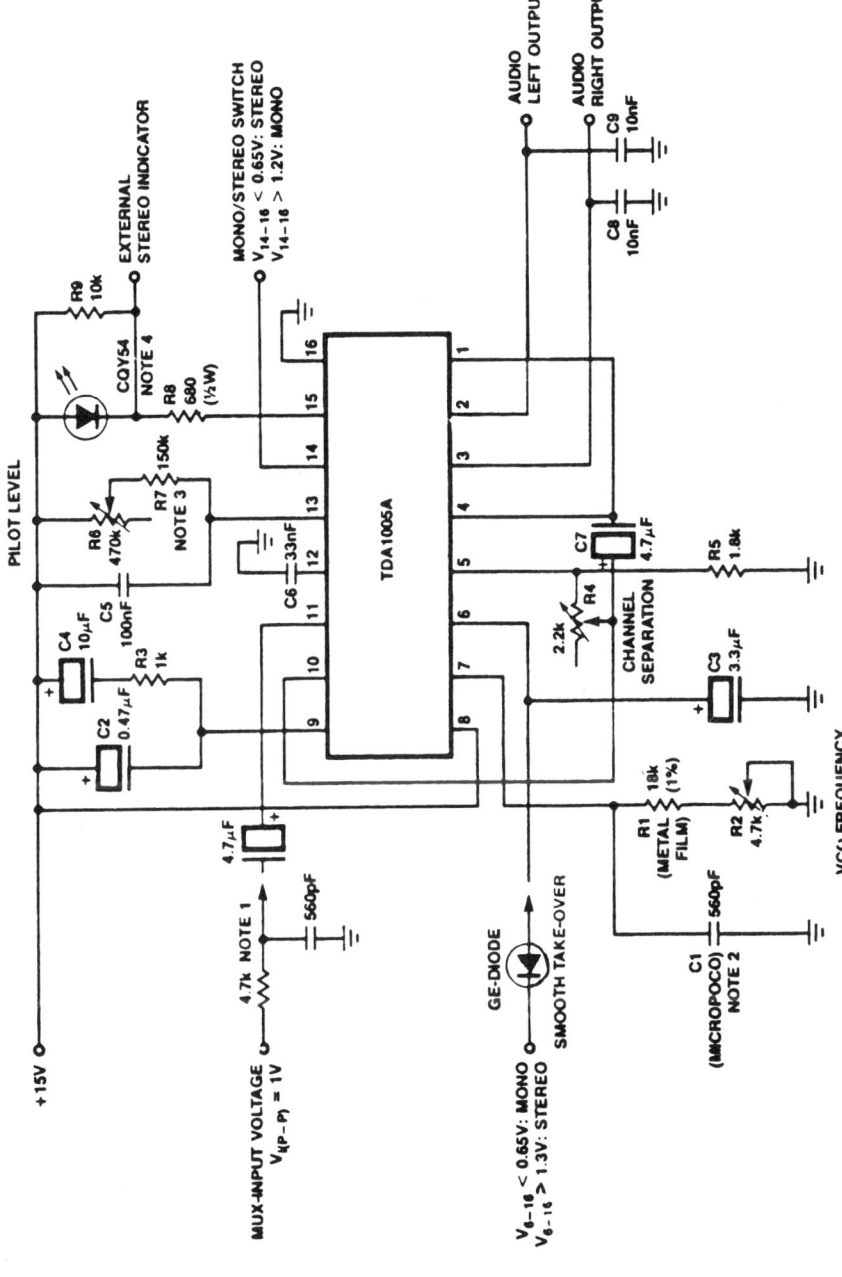

NOTES:
1. Micropoco capacitor has a temperature coefficient of $125.10^{-6} \pm 60.10^{-6} °C^{-1}$.
2. In simplified circuits a fixed resistor (e.g. 620k) can be used for a guaranteed switching level of $\leqslant 16mV$.
3. Either the LED circuit or an external stereo indicator can be used.

SIGNETICS

Fig. 3-5

9

SECOND-AUDIO PROGRAM ADAPTER

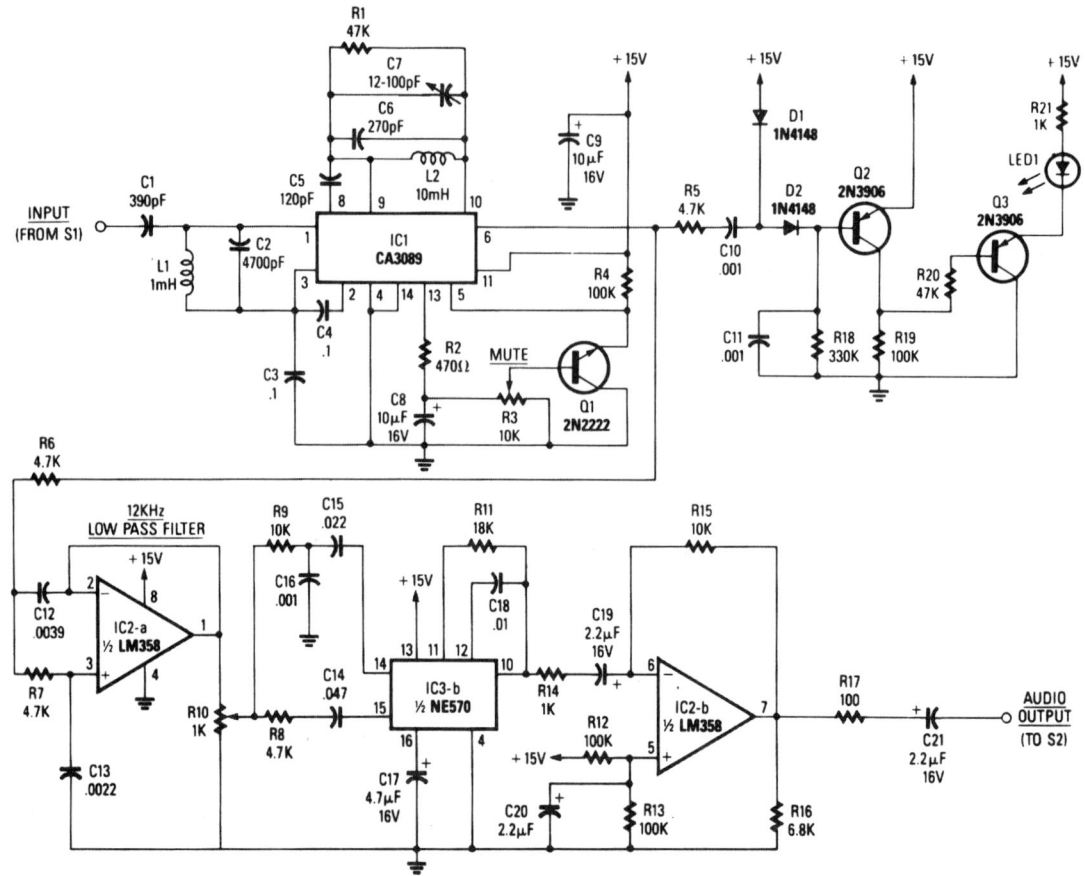

Reprinted with permission from Radio-Electronics Magazine 1989, R-E Experimenters Handbook.
Copyright Gernsback Publication, Inc., 1989.

Fig. 3-6

The baseband-audio input comes from the pole of switch S1 in the stereo decoder, and is coupled to IC1 (a CA3089) via a 78.6-kHz bandpass filter that consists of capacitors C1 and C2, and inductor L1. IC1 is a combination IF amplifier and quadrature detector normally used for FM radio systems operating within an IF of 10.7 MHz. The device works equally well at 78.6 kHz. Capacitors C6 and C7, and inductor L2 tune the detector section to 78.6 kHz, while C5 provides the necessary 90-degree phase shift for proper quadrature detector operation. The output voltage at pin 13 of IC1 is proportional to the level of the incoming signal. When the voltage at the wiper of potentiometer R3 reaches a predetermined threshold level, Q1 conducts, grounds pin 5 of IC1, and enables IC1's mute function.

Detected audio output from pin 6 of IC1 goes to IC2a, which is configured as a 12-kHz, −12 dB per octave, low-pass filter. The output of IC2a appears across potentiometer R10, which provides a means of adjusting the drive level into IC3b, the 2:1 compander.

Audio from the wiper of R10 is split into two paths: a high-pass filter (C14 and R8) provides a path to the rectifier input of the compander, and a bandpass filter (R9, C16, and C15) that feeds the audio input of the compander. A fixed 390-μs de-emphasis network is formed by C18 and R11 in conjunction with IC3b. Corrected audio at pin 10 of IC3b is coupled to IC2b, an output buffer amplifier.

Audio from pin 6 of IC1 is also coupled to an audio high-pass filter, R5 and C10, and fed to an audio rectifier, D1, D2, and C11. When an SAP signal is detected by IC1, it is rectified by D1 and D2; the resultant dc charges C11. An increasing positive voltage at the base of Q2 causes its current flow to decrease, so the voltage at Q2's collector also decreases. That, in turn, causes the base voltage of Q3 to drop, which causes Q3 to conduct, and thereby lights the LED.

TONE DECODER

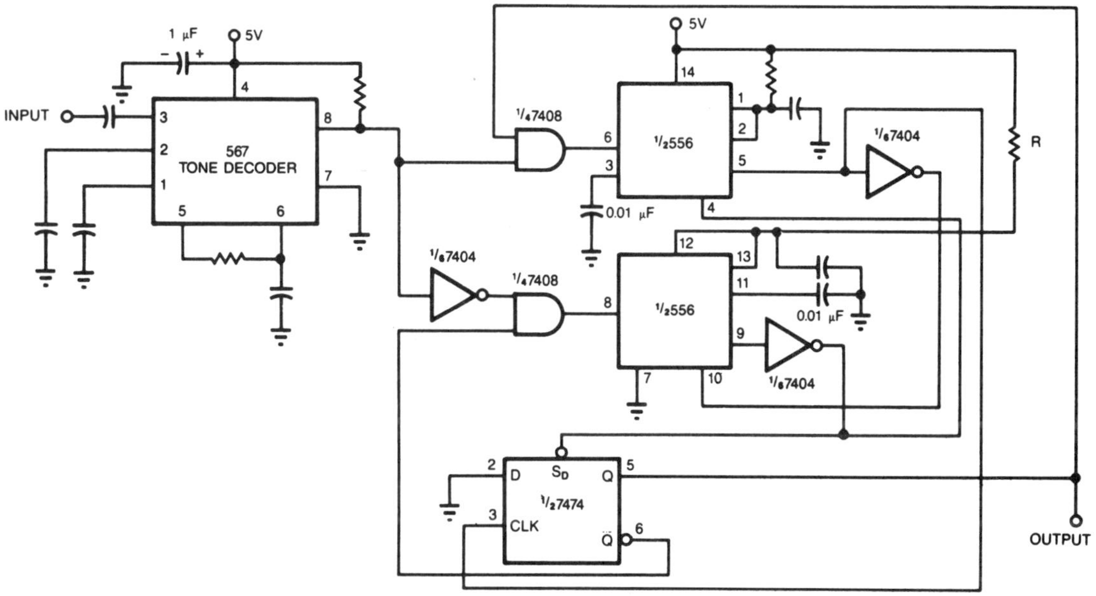

HANDS-ON ELECTRONICS/POPULAR ELECTRONICS *Fig. 3-7*

Adding a pair of one shots to the output of a 567 tone decoder renders it less sensitive to out-of-band signals and noise. Without the one shots, the 567 is prone to spurious output chatter. Other protection schemes, such as feeding back outputs or using an input filter, do not work as well as the one shots. The output of the 567 is high in the absence of a tone and becomes low when it detects a tone. The tone decoder triggers a one shot via an AND gate. The one shot's period is set to slightly less than the duration of a tone burst.

When the output of the tone decoder decreases, it triggers the second one shot. The second one shot's period is set to slightly less than the interval between tone bursts. The flip-flop enables and disables the inputs to one shots so that spurious outputs from the tone decoder do not affect the output.

ENCODER/DECODER

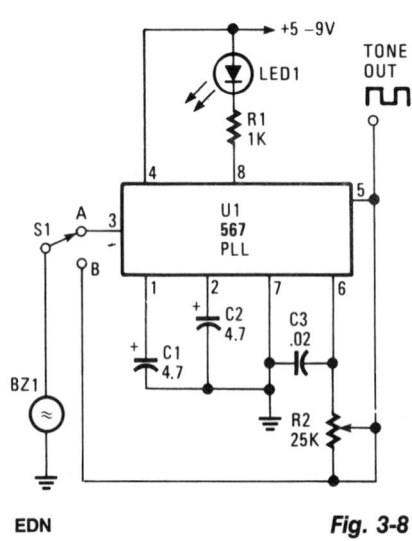

Fig. 3-8

The transducer circuit can be operated as either a tone encoder or decoder by changing the position of S1. The operating frequency of that dual-purpose circuit is determined by C3 and R2. Capacitors C1 and C2 are not critical and can be of almost any value between 1 and 5 mf. When the circuit is receiving an on-frequency signal, LED1 lights. Although a two wire piezo transducer with a resonance frequency of 2500 Hz was used in the circuit, any piezo unit should work—as long as the values of C3 and R2 are selected to tune to the transducer's operating frequency.

With power on and S1 in the *B* position, adjust R2 for the loudest tone output. The circuit should be tuned to the resonance frequency of the transducer. In that position, the circuit can be used as an acoustical or tone signal encoder. Next, switch to the *A* position and aim an on-frequency audible tone toward the transducer; the LED should light, indicating a decoded signal.

DIRECTION DETECTOR DECODER

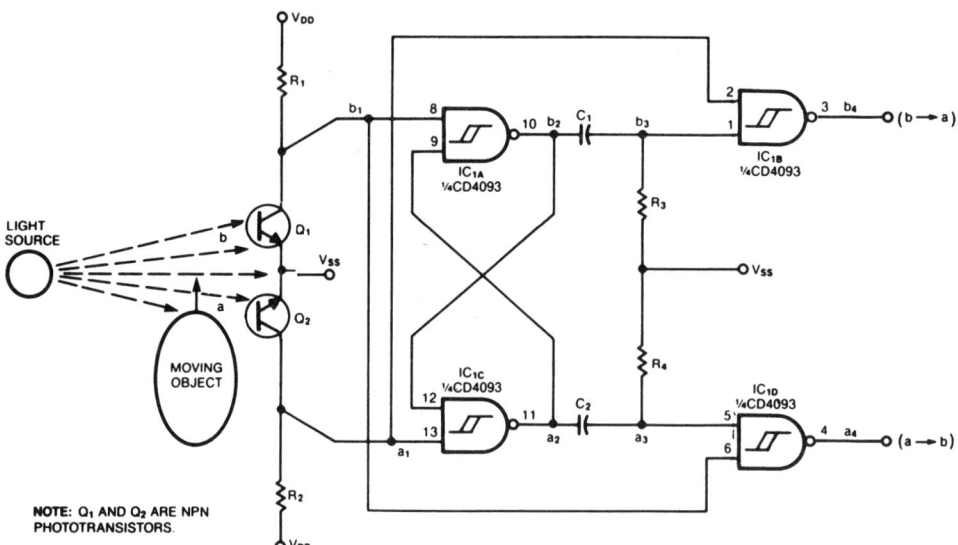

Fig. 3-9

DIRECTION DETECTOR DECODER *(Cont.)*

This circuit, which was developed to monitor the traffic of bumblebees in and out of the hive, differentiates a-to-b motion from b-to-a motion. When used with an optical decoder, the circuit distinguishes clockwise from counterclockwise rotation and provides a resolution of one output pulse per quadrature cycle.

Q1 and Q2 are mounted so that a moving object first blocks one phototransistor, then both, then the other. Depending on the direction in which the object is moving, either IC1B or IC1D emits a negative pulse when the moving object blocks the second sensor. An object can get as far as condition 3 and retreat without producing an output pulse; that is, the circuit ignores any probing or jittery motion. If an object gets as far as condition 4, however, a retreat will produce an opposite-direction pulse.

The time constants R3C1 and R4C2 set the output pulse width. A 100-kΩ/100-pF combination, for example, produces 10-μs pulses. Select a value for pullup resistors R1 and R2 from the 10 kΩ to 100 kΩ range, according to the sensitivity your application requires.

SOUND-ACTIVATED DECODER

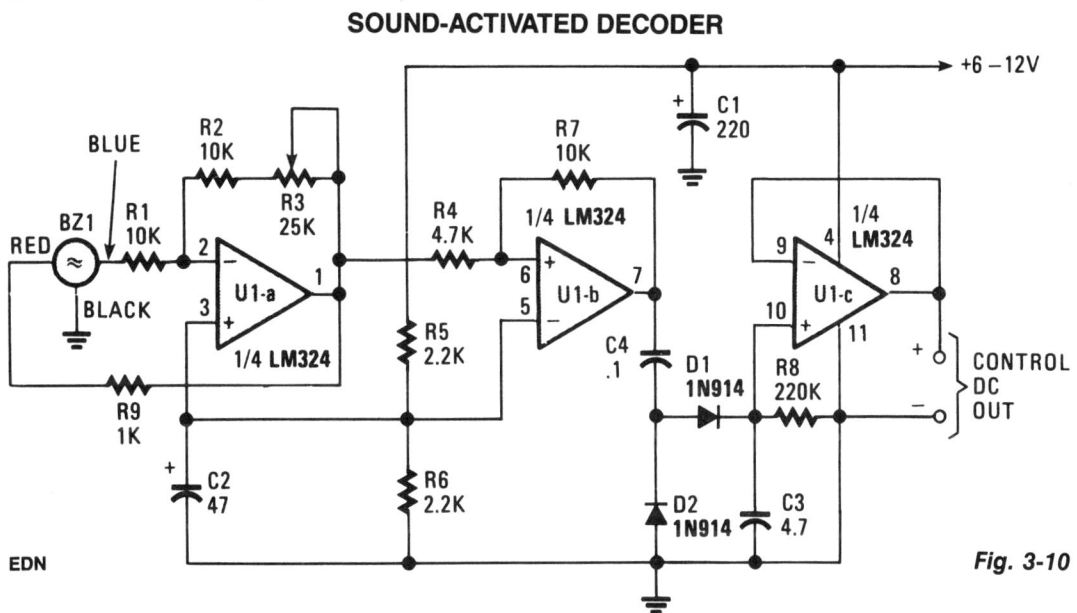

EDN

Fig. 3-10

The piezo transducer operates as a sound-pickup device as well as a frequency-selective filter. By controlling the gain of the op amps, the oscillator can be transformed into a sensitive and frequency-selective tone-decoder circuit. With the gain of U1a set just below the point of self oscillation, the transducer becomes sensitive to acoustically coupled audio tones that occur at or near its resonant frequency.

The circuit's output can be used to activate optocouplers, drive relay circuits, or to control almost any dc-operated circuit. The dc signal at the output of U1c varies with 0 to over 6 V, depending on the input-signal level. One unusual application for the sound-activated decoder would be in extremely high-noise environments, where normal broadband microphone pickup would be useless. Because piezo transducers respond only to frequencies within a very narrow bandwidth, little if any of the noise would get through the transducer.

FREQUENCY-DIVISION MULTIPLEX (FDM) STEREO DECODER

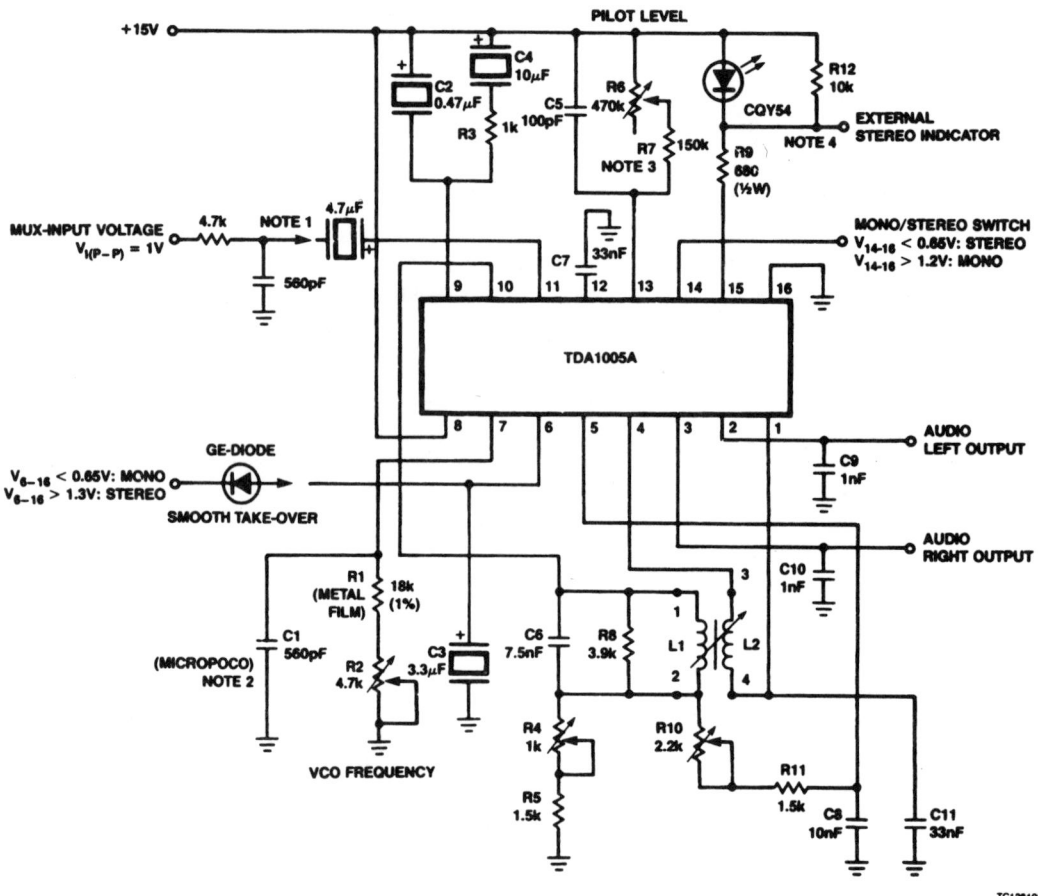

Coil data:
L_1 L_2 = 2.6mH
Q_{1-2} = 35; Q_{MIN} = 30
N_{1-2} = 357½ turns;
N_{3-4} = 297½ turns: scrambled wound with wire diameter 0.09mm, $\frac{E_{3-4}}{E_{1-2}} \times 100\% = 82\%$

NOTES:
1. The micropoco capacitor has a temperature coefficient of $125.10^{-6} \pm 60.10^{-6}$ k^{-1}.
2. In simplified circuits a fixed resistor (e.g. 620k) can be used for a guaranteed switching level of ≤ 16mV.
3. Either the LED circuit or an external stereo indicator can be used.

STEREO TV DECODER

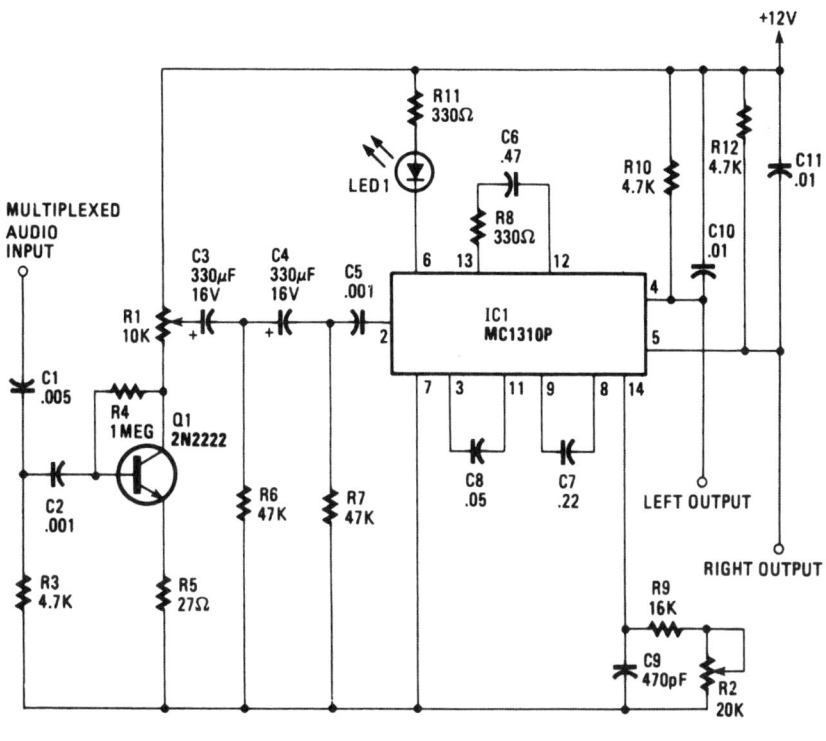

Fig. 3-12

The composite input signal is preamplified by transistor Q1 and is then coupled to the high-pass filter, composed of C3, C4, R6, and R7. The filtered audio is then passed to IC1, an MC1310P coilless stereo demodulator. That IC is normally used to demodulate broadcast-band FM signals, but by changing the frequency of its on-board VCO (voltage-controlled oscillator) slightly (from 19 kHz to 15.734 kHz), that IC can be used to detect stereo-TV signals.

Notice that the components connected to pin 14 control the VCO's frequency, hence the pilot-detect and carrier frequencies. For use in an FM receiver, the VCO would run at four times the 19-kHz pilot frequency (76 kHz), but for this application, it will run at four times the 15.734-kHz pilot frequency of stereo TV, 62.936 kHz. The MC1310P divides the master VCO signal by two in order to supply the 31.468 kHz carrier that is used to detect the L – R audio signal. The L – R signal undergoes normal FM detection, and at that point we've got two audio signals: L + R and L – R. The decoder block in the IC performs the addition and subtraction to produce the separate left and right signals. R10 and C10 form a de-emphasis network that compensates for the 75-μs pre-emphasis that the left channel underwent; R12 and C11 perform the same function for the right channel.

SCA (Background Music) DECODER

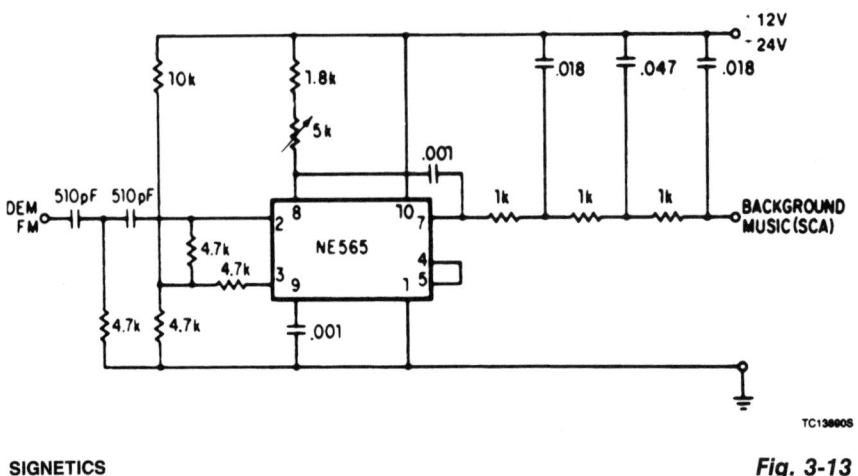

SIGNETICS

Fig. 3-13

A resistive voltage divider is used to establish a bias voltage for the input (Pins 2 and 3). The demodulated (multiplex) FM signal is fed to the input through a two-stage high-pass filter, both to effect capacitive coupling and to attenuate the strong signal of the regular channel. A total signal amplitude, between 80 and 300 mV, is required at the input. Its source should have an impedance of less than 10,000 Ω. The phase-locked loop is tuned to 67 kHz with a 5000-Ω potentiometer; only approximate tuning is required because the loop will seek the signal. The demodulated output (Pin 7) passes through a three-stage low-pass filter to provide de-emphasis and attenuate the high-frequency noise, which often accompanies SCA transmissions. Notice that no capacitor is provided directly at Pin 7; thus, the circuit is operating as a first-order loop. The demodulated output signal is in the order of 50 mV and the frequency response extends to 7 kHz.

10.8-MHz FSK DECODER

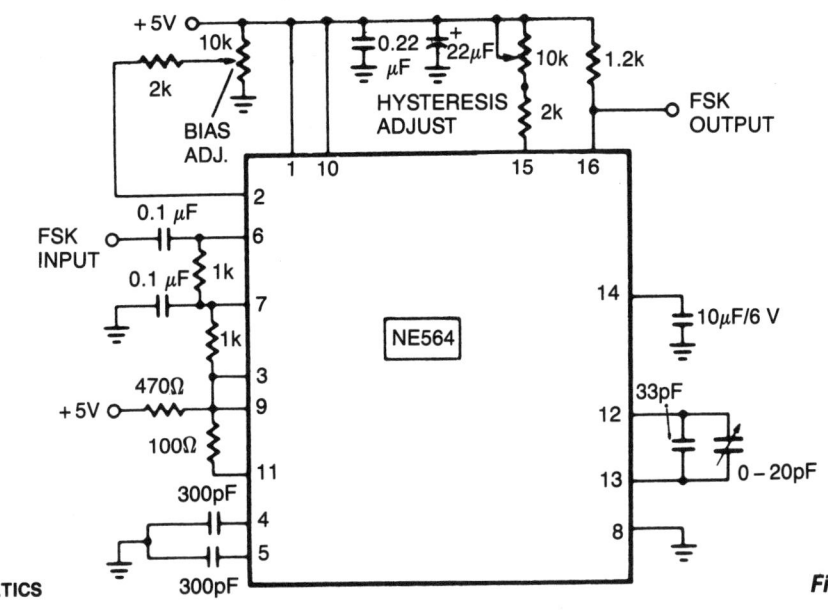

Fig. 3-14

DUAL-TIME CONSTANT TONE DECODER

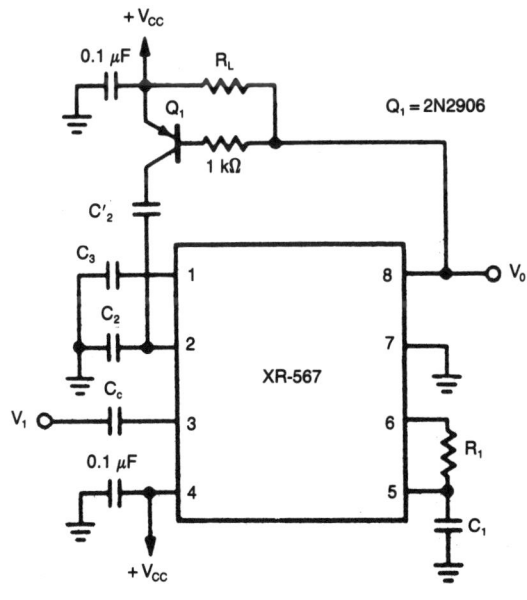

For some applications it is important to have a tone decoder with narrow bandwidth and fast response time. This can be accomplished by the dual-time constant tone decoder circuit shown. The circuit has two low-pass loop filter capacitors, C_2 and C'_2. With no input signal present, the output at pin 8 is high, transistor Q_1 is off, and C'_2 is switched out of the circuit. Thus, the loop low-pass filter is composed of C_2, which can be kept as small as possible for minimum response time. When an in-band signal is detected, the output at pin 8 will go low, Q_1 will turn on, and capacitor C'_2 will be switched in parallel with capacitor C_2. The low-pass filter capacitance will then be $C_2 + C'_2$. The value of C'_2 can be quite large in order to achieve narrow bandwidth. During the time that no input signal is being received, the bandwidth is determined by the capacitor C_2.

Fig. 3-15

4

Demodulators

The sources of the following circuits are contained in the Sources section, which begins on page 214. The figure number in the box of each circuit correlates to the source entry in the Sources section.

Narrow-Band FM Demodulator with Carrier
    Detect
Stereo Demodulator
AM Demodulator
FM Demodulator

NARROW-BAND FM DEMODULATOR WITH CARRIER DETECT

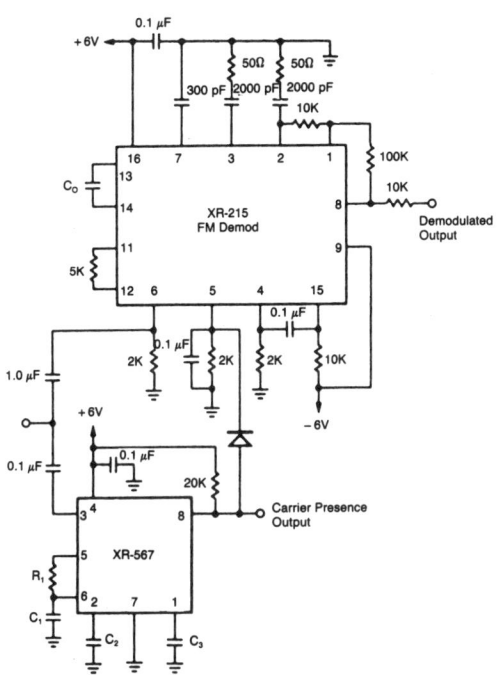

For FM demodulation applications where the bandwidth is less than 10% of the carrier frequency, an XR-567 can be used to detect the presence of the carrier signal. The output of the XR-567 is used to turn off the FM demodulator when no carrier is present, thus acting as a squelch. In the circuit shown, an XR-215 FM demodulator is used because of its wide dynamic range, high signal/noise ratio and low distortion. The XR-567 will detect the presence of a carrier at frequencies up to 500 kHz.

EXAR   *Fig. 4-1*

STEREO DEMODULATOR

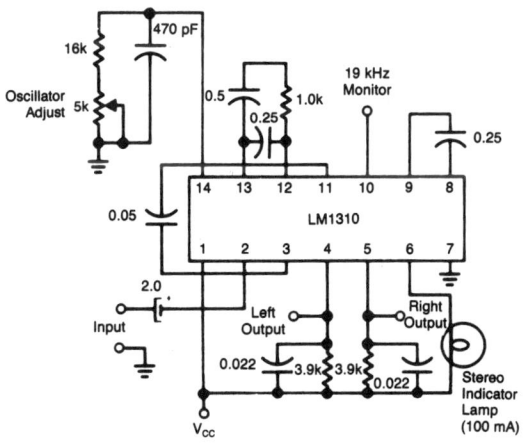

This circuit uses a single IC LM1310 to provide left and right outputs from a composite MPX stereo signal. Oscillator adjust R1 is set for 76 kHz (19 kHz at pin 10). C1 should be a silver mica or NPO ceramic capacitor.

NATIONAL SEMICONDUCTOR CORP.   *Fig. 4-2*

AM DEMODULATOR

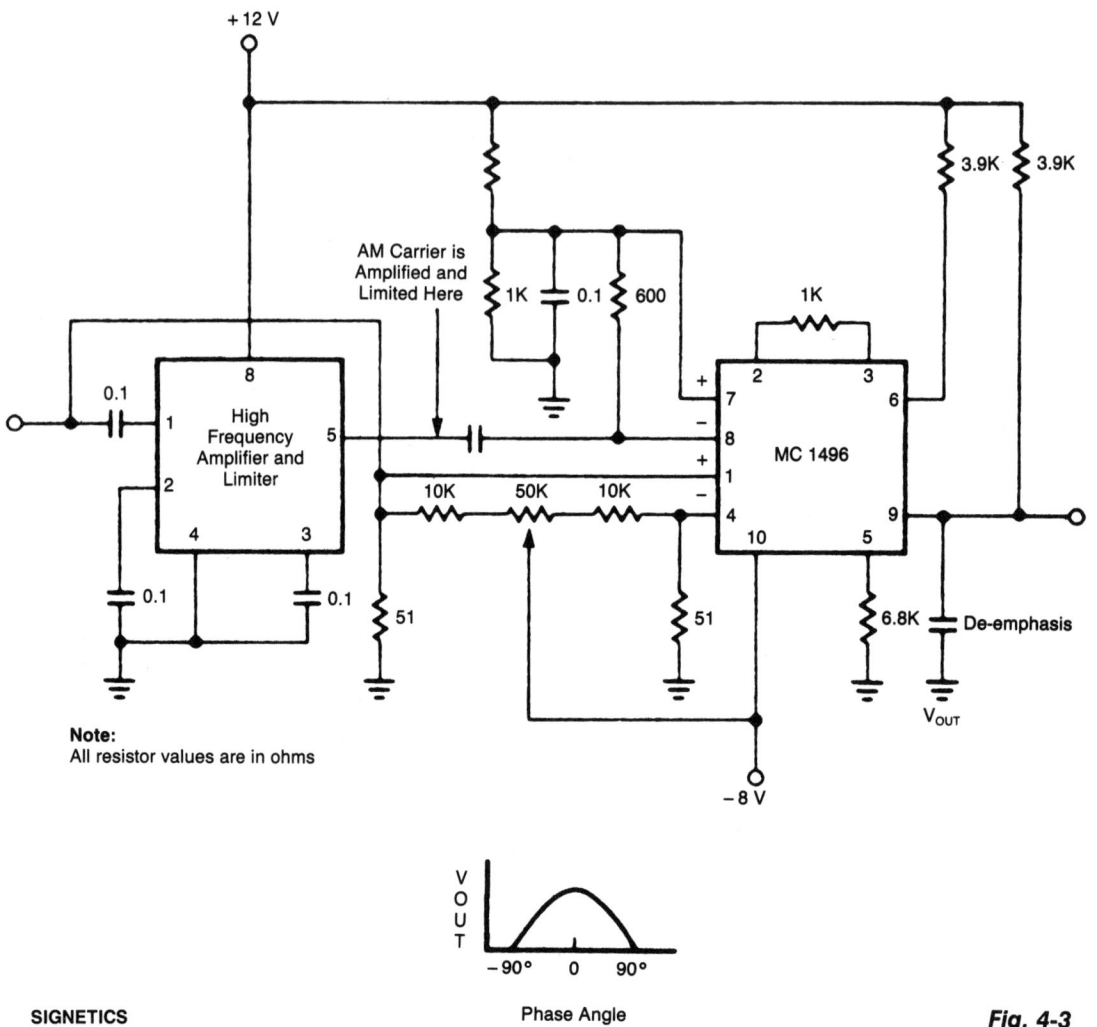

Note:
All resistor values are in ohms

SIGNETICS

Phase Angle

Fig. 4-3

Amplifying and limiting of the AM carrier is accomplished by the IF gain block, which provides 55 dB of gain or higher with a limiting of 40 μV. The limited carrier is then applied to the detector at the carrier ports to provide the desired switching function. The signal is then demodulated by the synchronous AM demodulator (1496), where the carrier frequency is attenuated as a result of the balanced nature of the device. Care must be taken not to overdrive the signal input so that distortion does not appear in the recorded audio. Maximum conversion gain is reached when the carrier signals are in phase, as indicated by the phase-gain relationship. Output filtering is also necessary to remove high-frequency sum components of the carrier from the audio signal.

FM DEMODULATOR

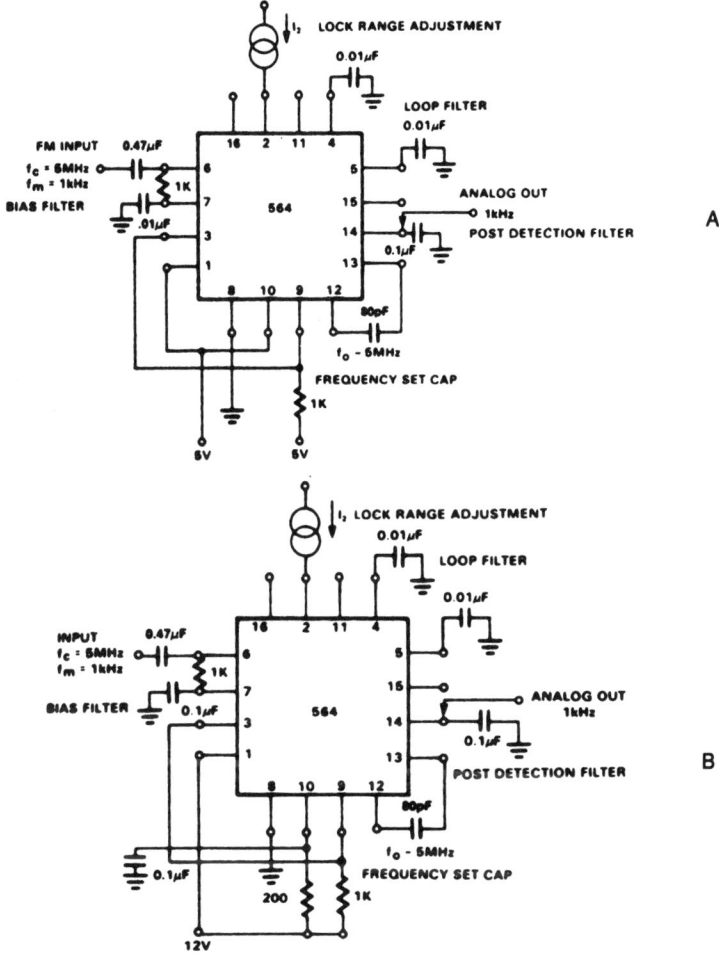

A

B

SIGNETICS

Fig. 4-4

The NE564 is used as an FM demodulator. The input signal is ac coupled with the output signal being extracted at Pin 14. Loop filtering is provided by the capacitors at Pins 4 and 5 with additional filtering being provided by the capacitor at Pin 14. Because the conversion gain of the VCO is not very high, to obtain sufficient demodulated output signal, the frequency deviation in the input signal should be 1% or higher.

5

Descramblers

The sources of the following circuits are contained in the Sources section, which begins on page 214. The figure number in the box of each circuit correlates to the source entry in the Sources section.

Sine-Wave Descrambler
Outband Descrambler
Gated Pulse Descrambler

SINE-WAVE DESCRAMBLER

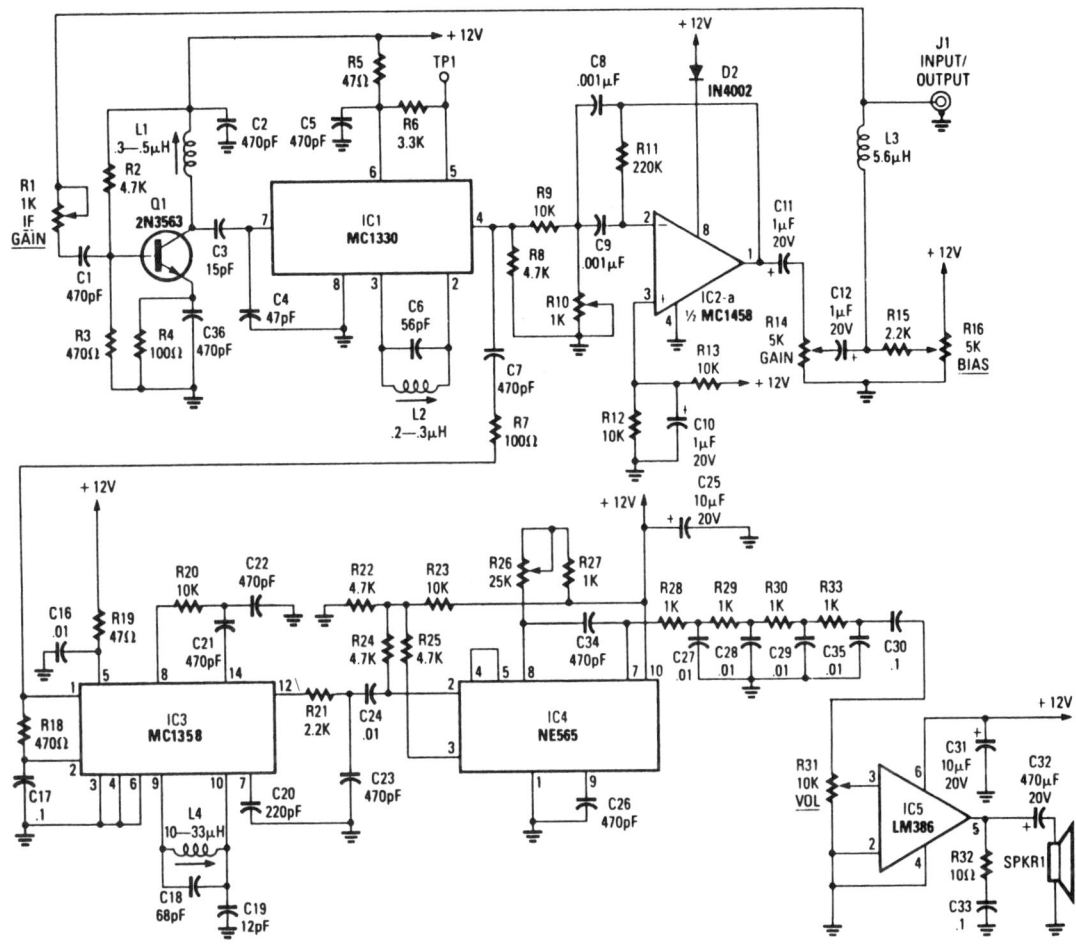

.—A COMPLETE SINEWAVE DESCRAMBLER. Easy to build, and relatively easy to align, this circuit completely removes the 15.75-kHz scrambling sinewave.

RADIO-ELECTRONICS

Fig. 5-1

This decoder features a sine-wave recovery channel and uses a PIN diode attenuator that is driven by the sine-wave recovery system to cancel out the sine-wave sync suppression signal. A kit is available from North Country Radio, P.O. Box 53, Wykagyl Station, New Rochelle, NY 10804.

OUTBAND DESCRAMBLER

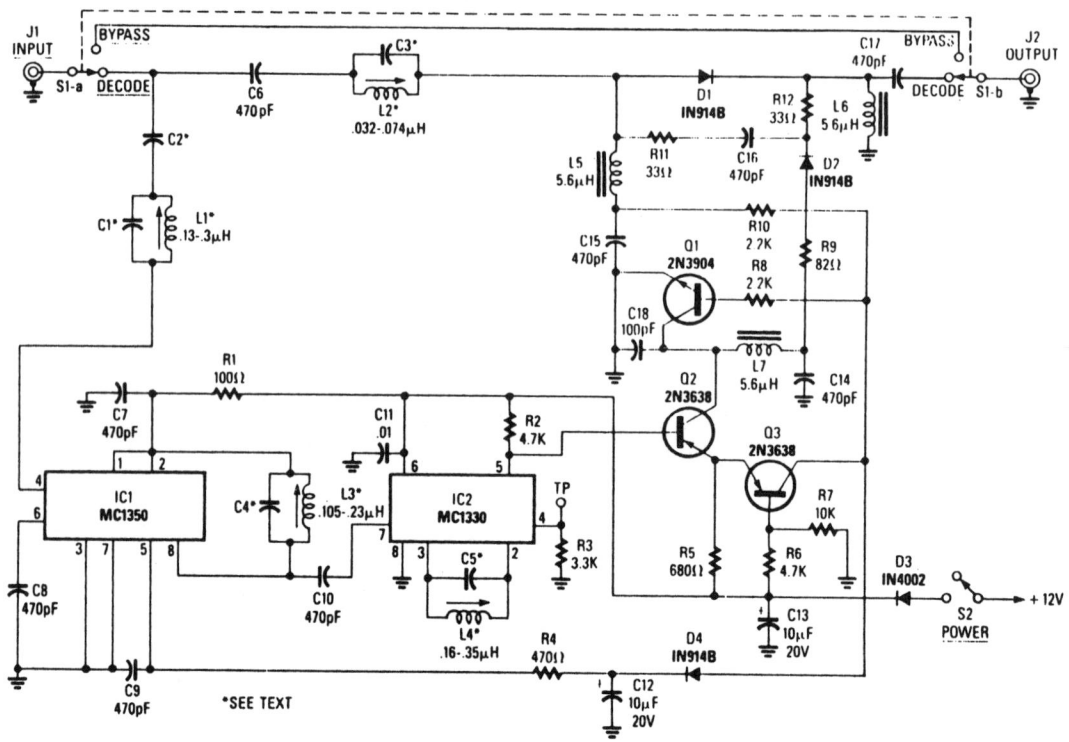

—FOR THE OUTBAND DECODER shown here to work, the cable company must provide at least a 1 millivolt signal. Values for C1–C5 and L1–L4 are found in Table 1.

TABLE 1—CAPACITOR AND COIL VALUES

	50 MHz	90–114 MHz
C1	5 pF	5 pF
C2	47 pF	12 pF
C3	200 pF	82 pF
C4	56 pF	12 pF
C5	56 pF	10 pF
L1	0.2 μH	0.2 μH
L2	0.05 μH	0.03 μH
L3	0.175 μH	0.2 μH
L4	0.175 μH	0.24 μH

Fig. 5-2

This circuit consists of an amplifier for the synch channel and a video detector, which controls an attenuator so that the gain of the systems is increased during synch intervals. A kit is available from North Country Radio, P.O. Box 53, Wykagyl Station, New Rochelle, NY 10804.

GATED PULSE DESCRAMBLER

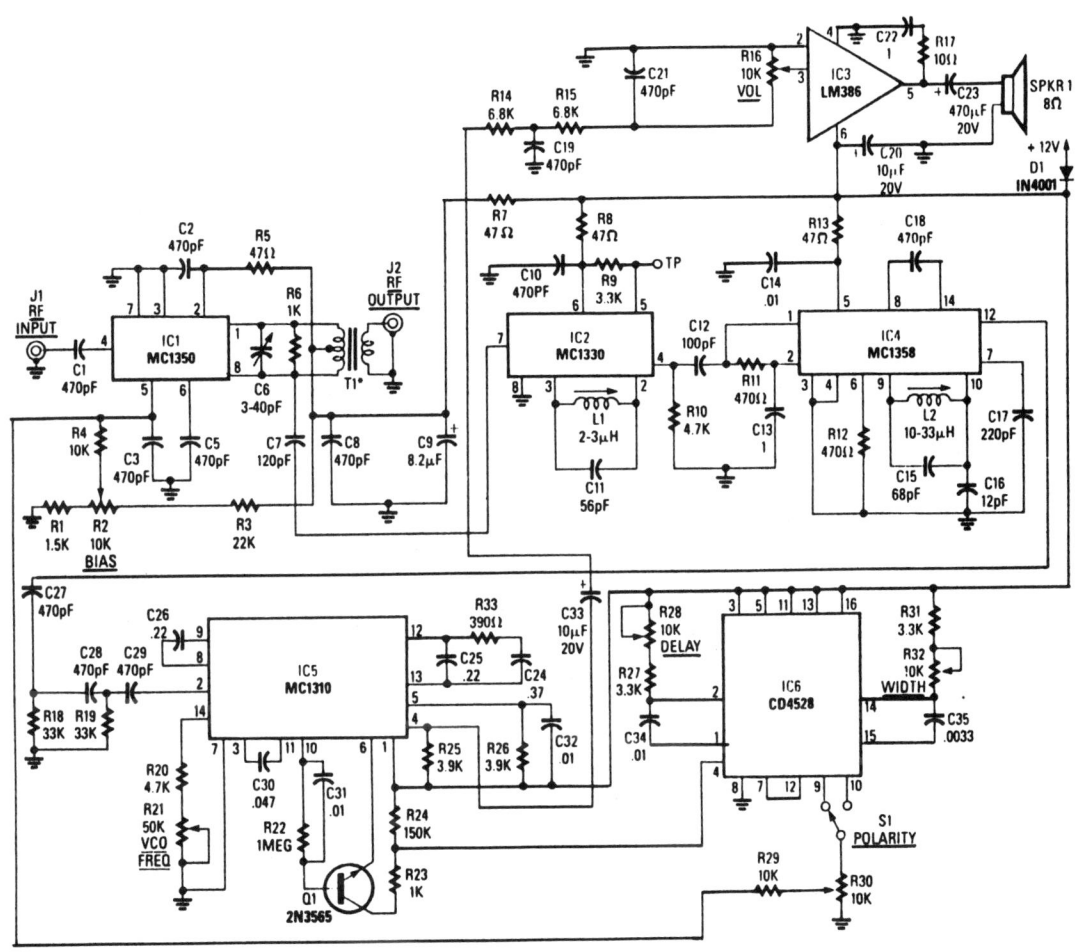

—DESCRAMBLE GATED-PULSE SIGNALS using this easy-to-build circuit. Information for winding transformer T1 and coil L1 can be found in the text.

Fig. 5-3

This circuit consists of an amplifier and video detector with a second subcarrier detector for synch-recovery purposes. A pulse-former circuit modulates the gain of the main channel increasing it during synch intervals. Provision for subcarrier audio descrambling is also provided. A kit is available from North Country Radio, P.O. Box 53, Wykagyl Station, New Rochelle, NY 10804.

6

Electrostatic Detector

The source of the following circuit is contained in the Sources section, which begins on page 214. The figure number contained in the box of the circuit correlates to the source entry in the Sources section.

Electrostatic Detector

ELECTROSTATIC DETECTOR

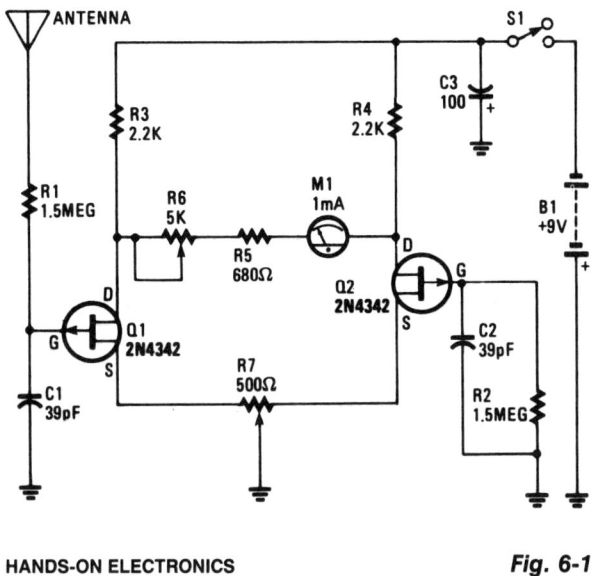

Fig. 6-1

In this electroscope, two junction FETs (Q1 and Q2) are connected in a balanced-bridge circuit. The gate input of Q1 is connected to the wire pick-up antenna, while Q2's gate is tied to the circuit's common ground through R2. That type of bridge circuit offers excellent temperature stability; therefore, Q1 is allowed to operate in an open-gate configuration. Potentiometer R7 is used to balance the bridge circuit, and R6 sets the maximum meter swing. Capacitors C1 and C2 help to reduce the 60-Hz pickup and add to the short-term stability of the circuit.

7

Flow Detectors

The sources of the following circuits are contained in the Sources section, which begins on page 214. The figure number in the box of each circuit correlates to the source entry in the Sources section.

Low Flow-Rate Thermal Flowmeter
Thermally Base Anemometer (Air Flowmeter)
Air-Flow Detector

LOW FLOW-RATE THERMAL FLOWMETER

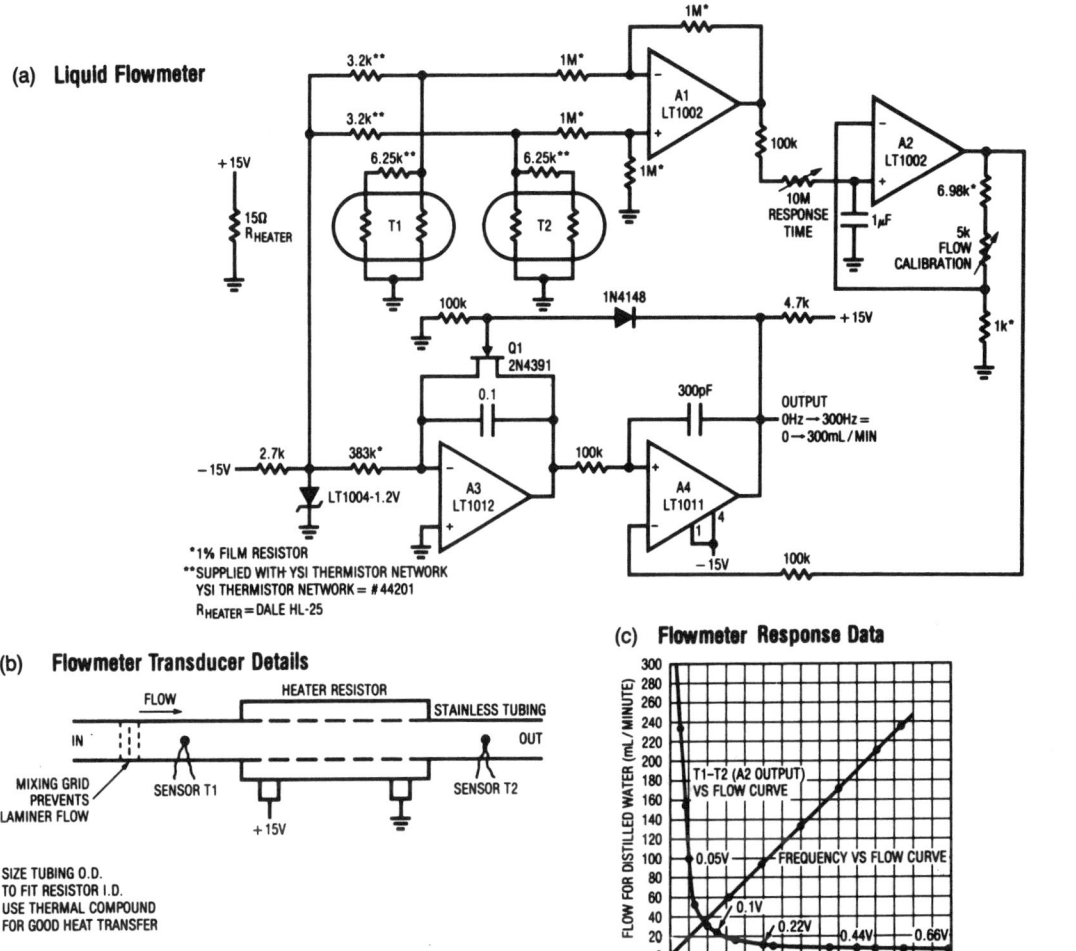

(a) Liquid Flowmeter

*1% FILM RESISTOR
**SUPPLIED WITH YSI THERMISTOR NETWORK
YSI THERMISTOR NETWORK = #44201
R HEATER = DALE HL-25

(b) Flowmeter Transducer Details

SIZE TUBING O.D.
TO FIT RESISTOR I.D.
USE THERMAL COMPOUND
FOR GOOD HEAT TRANSFER

LINEAR TECHNOLOGY CORP.

(c) Flowmeter Response Data

Fig. 7-1

This design measures the differential temperature between two sensors. Sensor T1, located before the heater resistor, assumes the fluid's temperature before it is heated by the resistor. Sensor T2 picks up the temperature rise induced into the fluid by the resistor's heating. The sensor's difference signal appears at A1's output. A2 amplifies this difference with a time constant set by the 10-MΩ adjustment. Figure 7-1c shows A2's output versus flow rate. The function has an inverse relationship. A3 and A4 linearize this relationship, while simultaneously providing a frequency output. A3 functions as an integrator that is biased from the LT1004 and the 383-kΩ input resistor. Its output is compared to A2's output at A4. Large inputs from A2 force the integrator to run for a long time before A4 can increase, turning on Q1 and resetting A3. For small inputs from A2, A3 does not have to integrate long before resetting action occurs. Thus, the configuration oscillates at a frequency which is inversely proportional to A2's output voltage. Because this voltage is inversely related to flow rate, the oscillation frequency linearly corresponds to flow rate.

THERMALLY BASED ANEMOMETER (AIR FLOWMETER)

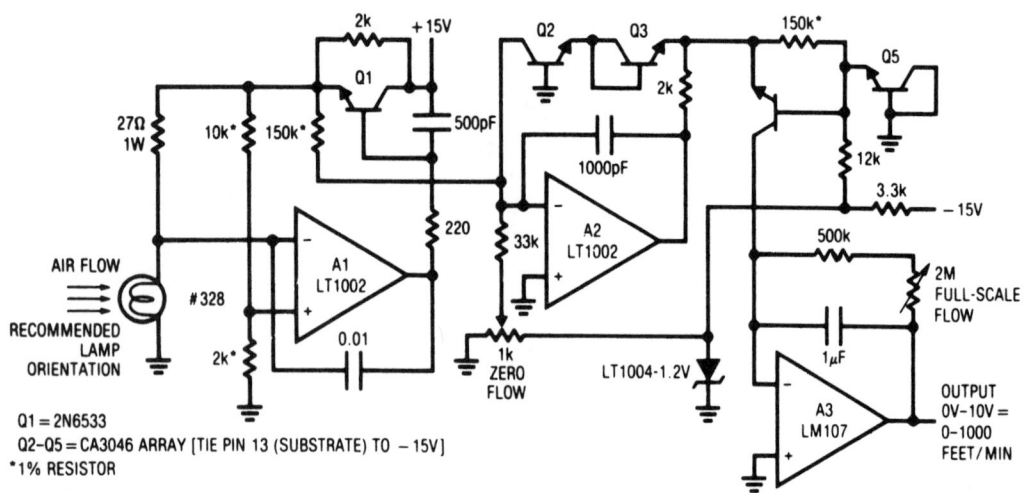

Q1 = 2N6533
Q2–Q5 = CA3046 ARRAY [TIE PIN 13 (SUBSTRATE) TO –15V]
*1% RESISTOR

LINEAR TECHNOLOGY CORP.

Fig. 7-2

This design used to measure air or gas flow works by measuring the energy required to maintain a heated resistance wire at constant temperature. The positive temperature coefficient of a small lamp, in combination with its ready availability, makes it a good sensor. A type 328 lamp is modified for this circuit by removing its glass envelope. The lamp is placed in a bridge, which is monitored by A1. A1's output is current amplified by Q1 and fed back to drive the bridge. When power is applied, the lamp is at a low resistance and Q1's emitter tries to come full on. As current flows through the lamp, its temperature quickly rises, and forces its resistance to increase. This action increases A1's negative input potential. Q1's emitter voltage decreases and the circuit finds a stable operating point. To keep the bridge balanced, A1 acts to force the lamp's resistance, hence its temperature, constant. The 20-kΩ to 2-kΩ bridge values have been chosen so that the lamp operates just below the incandescence point.

To use this circuit, place the lamp in the air flow so that its filament is at a 90° angle to the flow. Next, either shut off the air flow or shield the lamp from it and adjust the zero-flow potentiometer for a circuit output of 0 V. Then, expose the lamp to air flow of 1000 feet/minute and trim the full flow potentiometer for 10-V output. Repeat these adjustments until both points are fixed. With this procedure completed, the air flowmeter is accurate within 3% over the entire 0- to 1000-foot/minute range.

AIR-FLOW DETECTOR

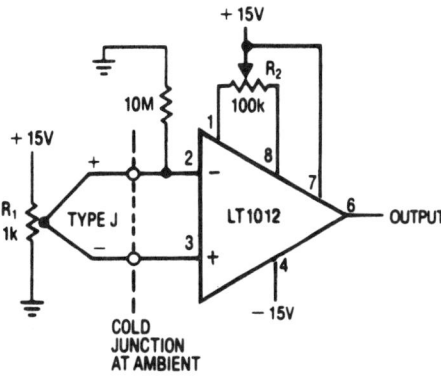

Mount R_1 in airflow.
Adjust R_2 so output goes high when airflow stops.

LINEAR TECHNOLOGY CORP.   *Fig. 7-3*

8

Gas/Smoke/Vapor Detector

The sources of the following circuits are contained in the Sources section, which begins on page 214. The figure number in the box of each circuit correlates to the source entry in the Sources section.

Gas and Smoke Detector
Ionization Chamber Smoke Detector
Gas Analyzer
Toxic Gas Detector
Gas and Vapor Detector
Furnace Exhaust Gas-Temperature Monitor with Low Supply Detection
Methane-Concentration Detector with Linearized Output
Smoke/Gas/Vapor Detector
Gas/Smoke Detector I

Smoke Detector I
SCR Smoke Alarm
Gas/Smoke Detector II
Smoke Detector II
Ionization-Chamber Smoke Detector
Photoelectric Smoke Detector (Nonlatching)
1.9-V Battery-Operated Ionization-Type Smoke Detector
Line-Operated Photoelectric Smoke Alarm
Smoke Detector III

GAS AND SMOKE DETECTOR

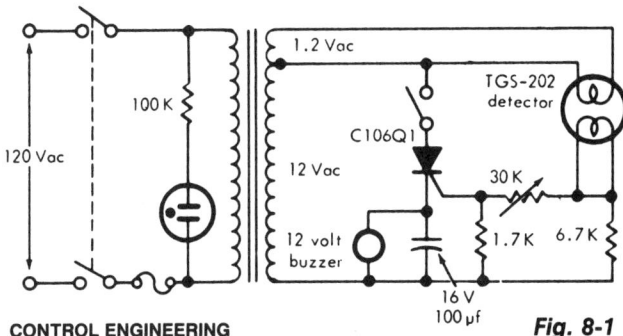

CONTROL ENGINEERING

Fig. 8-1

This circuit can detect smoke and a number of gases (CO, CO_2, methane, coal gas, and others) with a 10-ppm sensitivity. It uses a heated-surface semiconductor sensor. Detection occurs when the gas concentration increase causes a decrease of the sensor element internal resistance. The switch in series with the SCR is used to reset the alarm.

IONIZATION-CHAMBER SMOKE DETECTOR

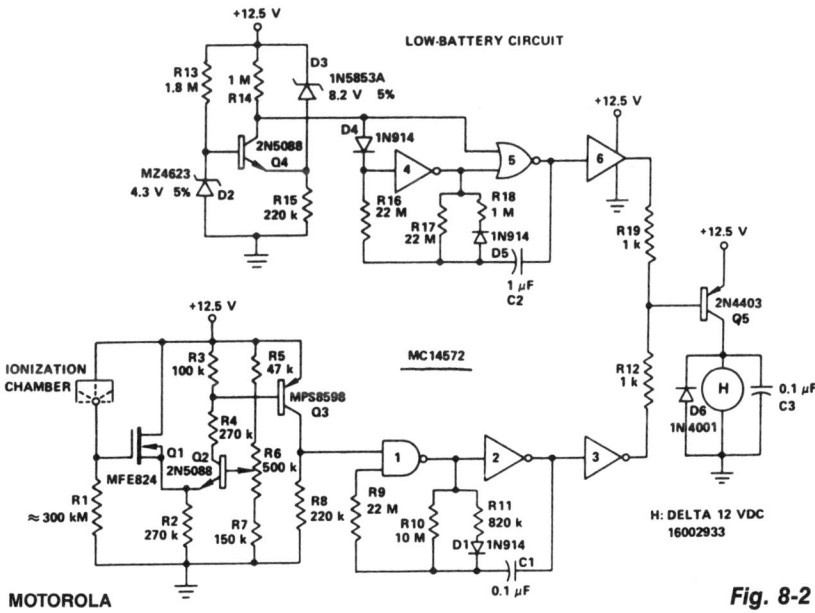

MOTOROLA

Fig. 8-2

This battery-operated ionization-chamber smoke detector includes a circuit to generate a unique alarm when the battery reaches the end of its useful life. The circuit uses the MCMOS MC14572 for two alarm oscillators (smoke and low battery). This circuit additionally uses five discrete transistors as buffers and comparators.

GAS ANALYZER

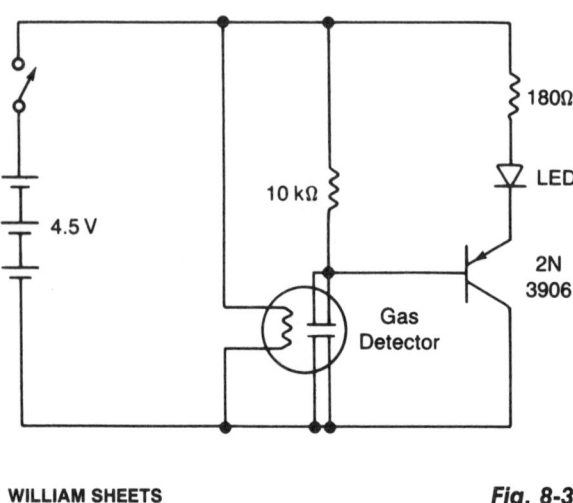

180Ω

LED

10 kΩ

4.5 V

2N
3906

Gas
Detector

Fig. 8-3

The circuit shows a simple yes/no gas detector. Three 1.5-V D cells are used as a power supply, with S1 acting as an on/off switch. The heater is energized directly from the battery, while the electrodes are in series with a 10-kΩ resistor. The voltage across this resistor is monitored by a pnp transistor. When the sensor is in clean air, the resistance between the electrodes is about 40 kΩ, so only about 0.9 V is dropped across the 10-kΩ resistor. This voltage is insufficient to turn on the transistor, because of the extra 1.6 V, which is required to forward bias the LED in series with the emitter. When the sensor comes in contact with contaminated air, the resistance starts to fall, increasing the voltage dropped across the 10-kΩ resistor. When the sensor resistance falls to about 10 kΩ or less, the transistor starts to turn on, current passes through the LED, and causes it to emit. The 180-Ω resistor limits the current through the LED to a safe value.

TOXIC GAS DETECTOR

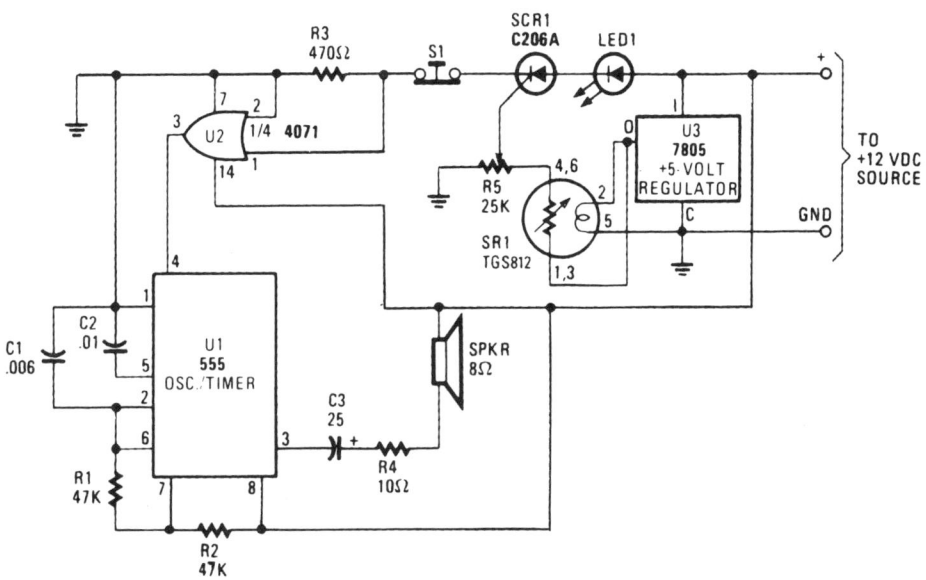

Fig. 8-4

The major device in the circuit is SR1 (a TGS812 toxic-gas sensor, which is manufactured by Figaro Engineering Inc.). The gas-sensitive semiconductor (acting like a variable resistor in the presence of toxic gas) decreases in electrical resistance when gaseous toxins are absorbed from the sensor surface. A 25-kΩ potentiometer (R5) connected to the sensor serves as a load, voltage-dividing network, and sensitivity control and has its center tap connected to the gate of SCR1. When toxic fumes contact the sensor, which decreases its electrical resistance, current flows through the load (potentiometer R5). The voltage developed across the wiper of R5, which is connected to the gate of SCR1, triggers the SCR into conduction. A 7805 regulator is used to meet the 5-V requirement for the heater and semiconductor elements.

GAS AND VAPOR DETECTOR

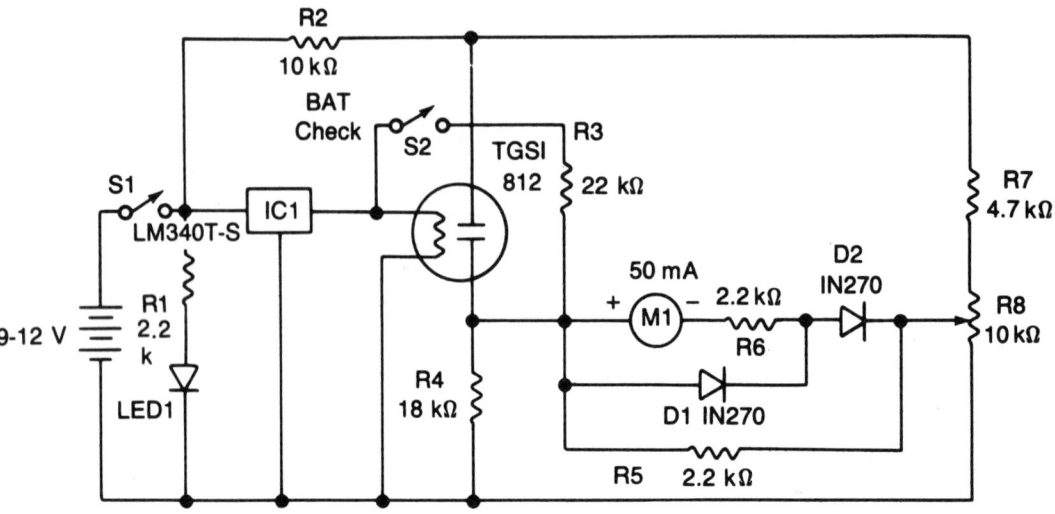

WILLIAM SHEETS

Fig. 8-5

The power drain is approximately 150 mA. IC1 provides a regulated 5-V supply for the filament heater of the sensor. The gas-sensitive element is connected as one arm of a resistance bridge consisting of R4, R7, R8 and the meter M1 with its associated resistors R5 and R6. The bridge can be balanced by adjusting R8 so that no current flows through the meter. A change in the sensor's resistance, caused by detection of noxious gases, will unbalance the circuit and deflect the meter. Diodes D1 and D2, and resistor R5 protect the meter from overload, while R6 determines overall sensitivity. R2 limits the current through the sensor; R1.and LED1 indicate that the circuit is working so that you do not drain the battery leaving the unit on inadvertently; R3 and S2 give a battery level check.

FURNACE-EXHAUST GAS-TEMPERATURE
MONITOR WITH LOW SUPPLY DETECTION

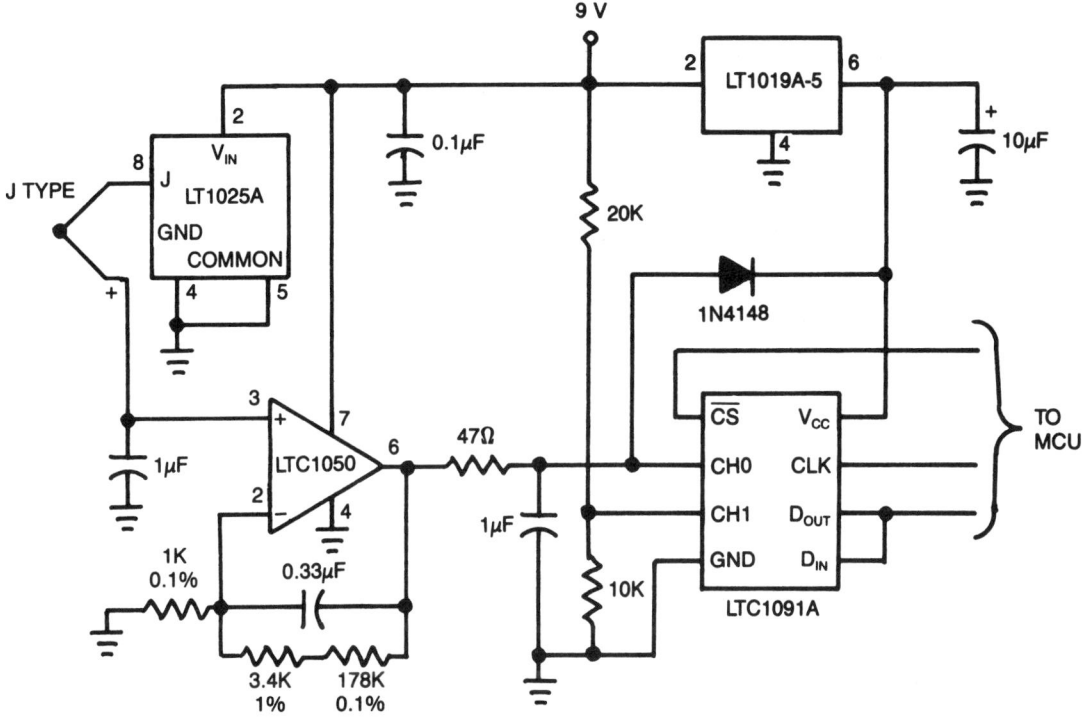

LINEAR TECHNOLOGY

Fig. 8-6

This circuit can be used to measure exhaust gas temperature in a furnace. The 10-bit LTC1091A gives 0.5°C resolution over a 0°C to 500°C range. The LTC1050 amplifies and filters the thermocouple signal, the LT1025A provides cold-junction compensation and the LT1019A provides an accurate reference. The J-type thermocouple characteristic is linearized digitally inside the MCU. Linear interpolation between known temperature points spaced 30°C apart introduces less than 0.1°C error. The 20-kΩ/10-kΩ divider on CH1 of the LTC1091 provides low supply voltage detection. Remote location is easy, with data transferred from the MCU to the LTC1091 via the three-wire serial port.

METHANE-CONCENTRATION DETECTOR WITH LINEARIZED OUTPUT

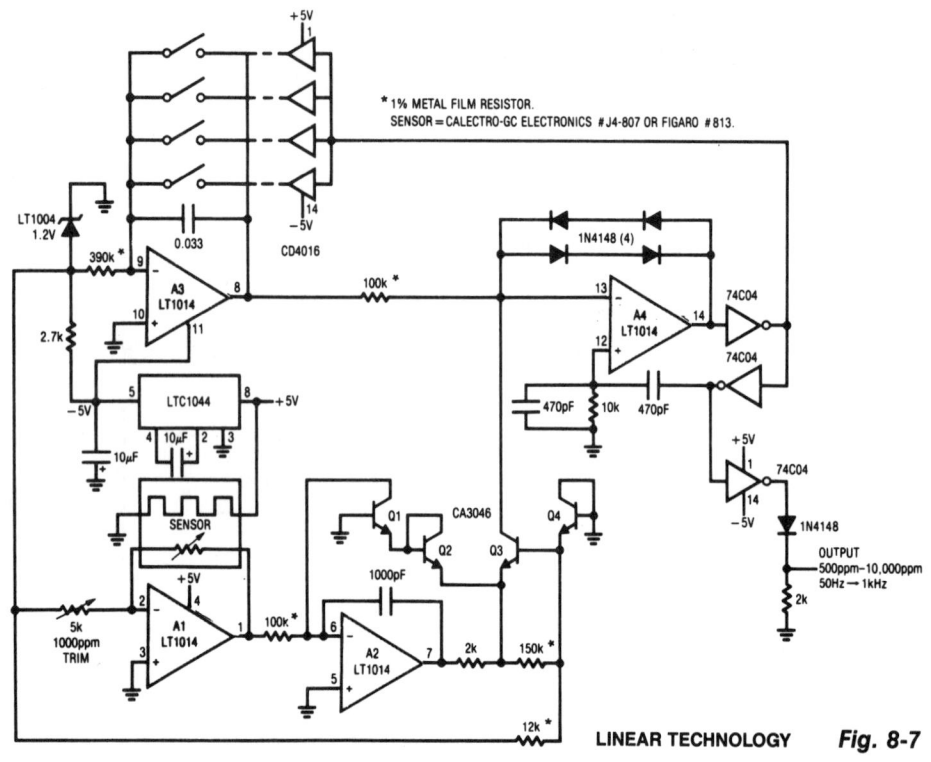

* 1% METAL FILM RESISTOR.
SENSOR = CALECTRO-GC ELECTRONICS #J4-807 OR FIGARO #813.

LINEAR TECHNOLOGY　　　*Fig. 8-7*

SMOKE/GAS/VAPOR DETECTOR

PARTS LIST

V_B	9 Volt Transistor Battery	R3	4.7, $\frac{1}{2}$W Resistor
C1	47μF, 35 V	R4	5 Ohms, 12 Watt
C2, C3	100 pf, 600V	R5	68k $\frac{1}{2}$W Resistor
C4	0.01μF, 50V	R6	1.3k $\frac{1}{2}$W Resistor
D1	1N662 Silicon Diode	R7	Fenwal GB32J2 or Veco 35D6
D2	1N750 Zener Diode, 4.7V	S1	Toggle Switch SPST.
D3, D4, D5	1N4002 Diode	F1	Fuse, 6/10 A Slow-Blow, Littlefuse A/3AG/MDL
Q1	2N1132 PNP Transistor	T1	Transformer 117 VAC 60 Hz Primary
Q2	2N4096 SCR		12.6V CT @ 1.2 A Secondary
R1	25k Potentiometer Ohmite Style 53C1		Allied Electronics 6K94HF
R2	5.1k, $\frac{1}{2}$W Resistor	TGS	Sensor TGS-202

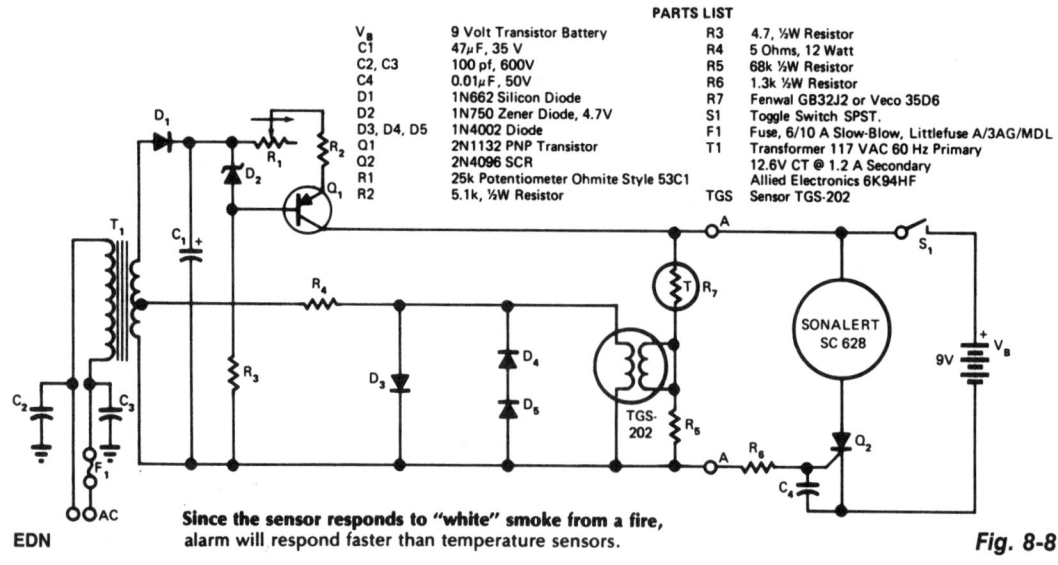

Since the sensor responds to "white" smoke from a fire, alarm will respond faster than temperature sensors.

EDN　　　*Fig. 8-8*

Transformer T1 supplies power to the heater of the sensor. Because the sensor is fairly sensitive to heater voltage, diodes D3, D4, and D5 regulate the heater voltage. T1, together with D1 and C2, forms a dc power supply, whose current is regulated by Q1 and adjusted by R1. The constant current from Q1 feeds a variable resistance, consisting of thermistor R7 and the parallel combination of R5 and the sensor resistance. When a hazard causes the voltage at A-A to drop, the net voltage at the SCR gate turns positive, triggers the SCR on, and operates the alarm. The alarm draws a small amount of current, so the battery will last a long time. Switch S1 turns off the alarm and resets the SCR.

GAS/SMOKE DETECTOR I

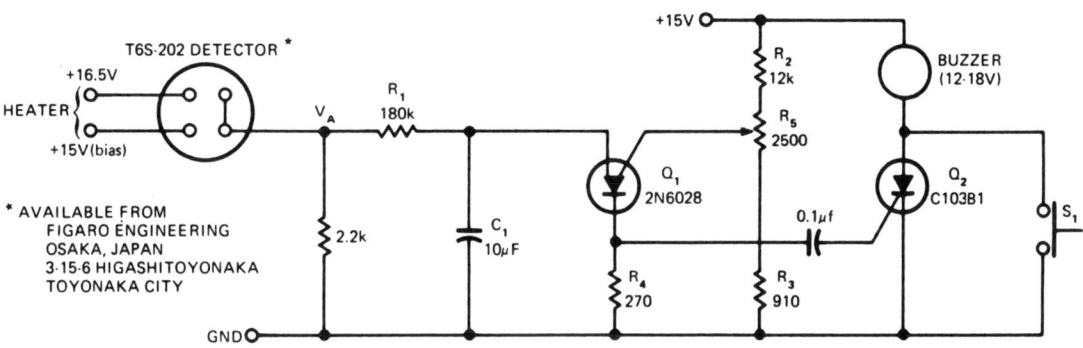

EDN

Fig. 8-9

The sensor is based on the selective absorption of hydrocarbons by an n-type metal-oxide surface. The heater in the device serves to burn off the hydrocarbons once smoke or gas is no longer present in the immediate area; hence, the device is reusable. When initially turned on, a 15-minute warm-up period is required to reach equilibrium ($V_A \cong 0.6$ V) in a hydrocarbon-free environment. When gas or smoke is introduced near the sensor, V_A will quickly rise (rate and final equilibrium depend on the type of gas and concentration) and trigger Q1, a programmable unijunction transistor. The voltage pulse generated across R4 triggers Q2, sounding the buzzer until S1 resets the unit. R1 and C1 give a time delay to prevent small transient waves of smoke, such as from a cigarette, from triggering the alarm. Triggering threshold is set by R5, R2, and R3; with the components shown, between 50 and 200 ppm of hydrocarbons can be easily detected. Because it is somewhat sensitive to heater voltage, a regulated supply should be used. Power requirements are 1.5 V at 500 mA for the heater and 15 V at 30 mA, depending on the type of buzzer for the bias supply.

SMOKE DETECTOR I

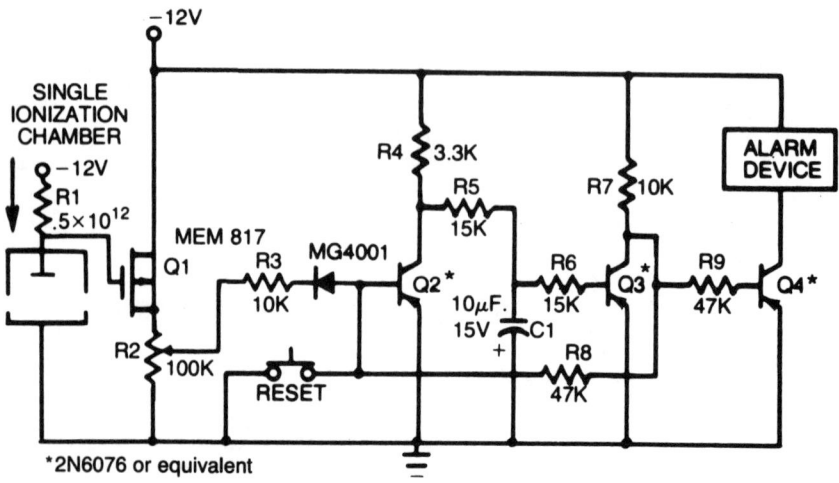

GENERAL INSTRUMENT MICROELECTRONICS

Fig. 8-10

This smoke detector uses a MEM 817 p-channel enhancement-mode MOSFET as its buffer amplifier. Operation of the sensor is based on a decrease in the current when smoke enters the chamber, thereby causing a negative voltage excursion at the gate of the buffer MOSFET. Quiescent voltage values at the output of the chamber vary from about −4 V to −6 V, and detection of smoke will result in an excursion of about −4 V. The MOSFET is connected as a source follower.

SCR SMOKE ALARM

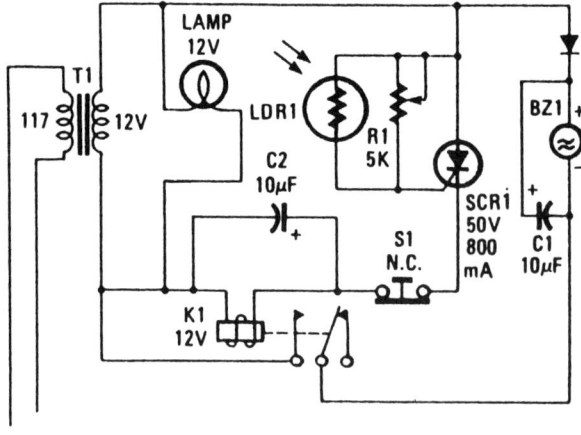

POPULAR ELECTRONICS

Fig. 8-11

GAS/SMOKE DETECTOR II

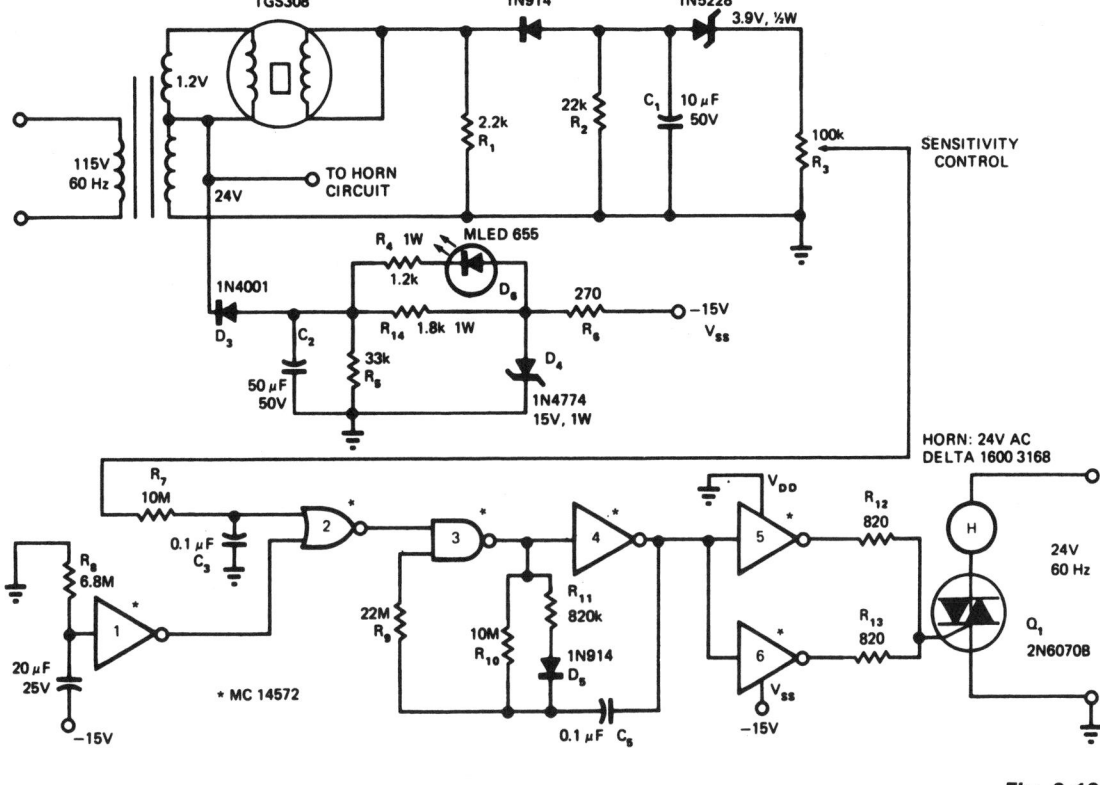

EDN

Fig. 8-12

In the presence of smoke or gas, the ac output voltage increases and becomes rectified, filtered and zener-diode coupled (D2 for thresholding) to sensitivity control R3. Under no gas condition, the output equals approximately 0 V (high). When gas is present, the output will be a negative value (low) sufficient to overcome the threshold McMOS gate 2 and D2. The circuit shown uses a TGS 308 sensor, a general-purpose gas detector that is not sensitive to smoke or carbon monoxide. If smoke is the primary element to be detected, use the TGS 202 sensor. The two sensors are basically identical; the main differences lie in the heater voltage and the required warm up time delay. The TGS requires a 1.2-V heater and a 2-minute delay, whereas the TGS 202 requires 1.5 V and 5 minutes, respectively.

The system uses a McMOS gated oscillator directly interfacing with a triac-controlled ac horn. Using the MC14572 HEX functional gate, four inverters, one two-input NAND gate and one two-input NOR gate, the circuit provides the complete gas/smoke detector logic functions time delay, gated astable multivibrator control and buffers operation. The 24-Vac horn produces an 85-/90-dB sound level output at a distance of 10 ft. Controlled by the astable multivibrator, the horn generates a pulsating alarm—a signal that might be advantageous over a continuous one in some noise environments.

SMOKE DETECTOR II

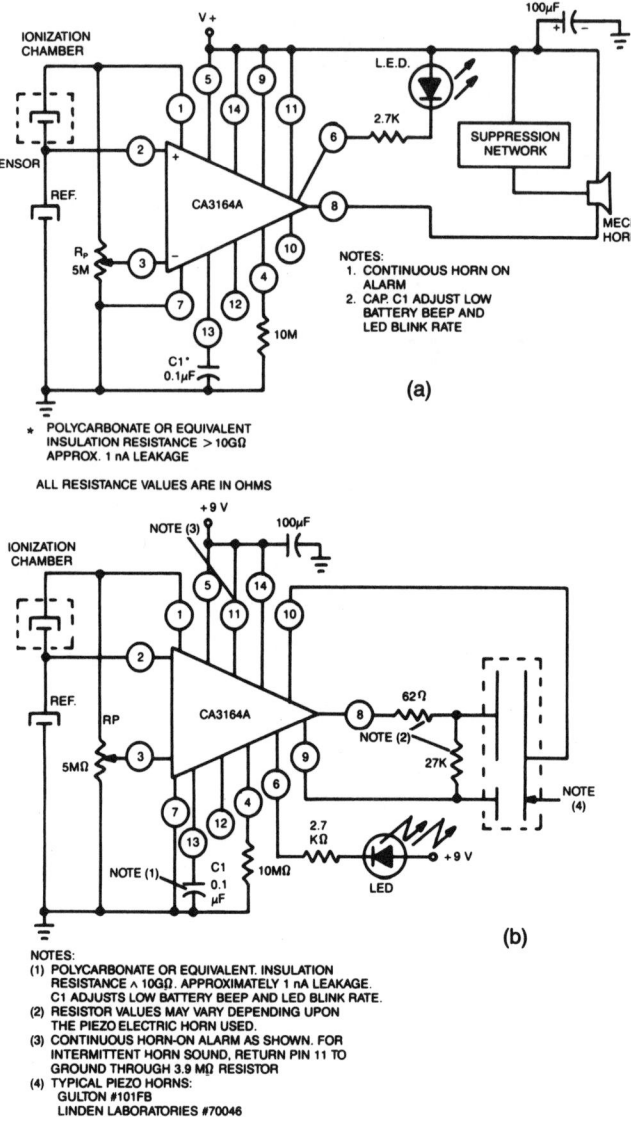

Fig. 8-13

GE/RCA

This smoke detector uses a CA3164A BiMOS detector/alarm system. For operation as smoke detector with electromechanical horn (Fig. 8-13a), the output of driver at terminal 8 is used. Large npn transistor Q3, with an active pull-up, and transistor Q2 provide over 300 mA of drive current. For operation as a smoke detector with a piezoelectric horn (Fig. 8-13b), the circuit requires output from a second inverting amplifier at terminal 10, as well as the output from terminal 8.

IONIZATION-CHAMBER SMOKE DETECTOR

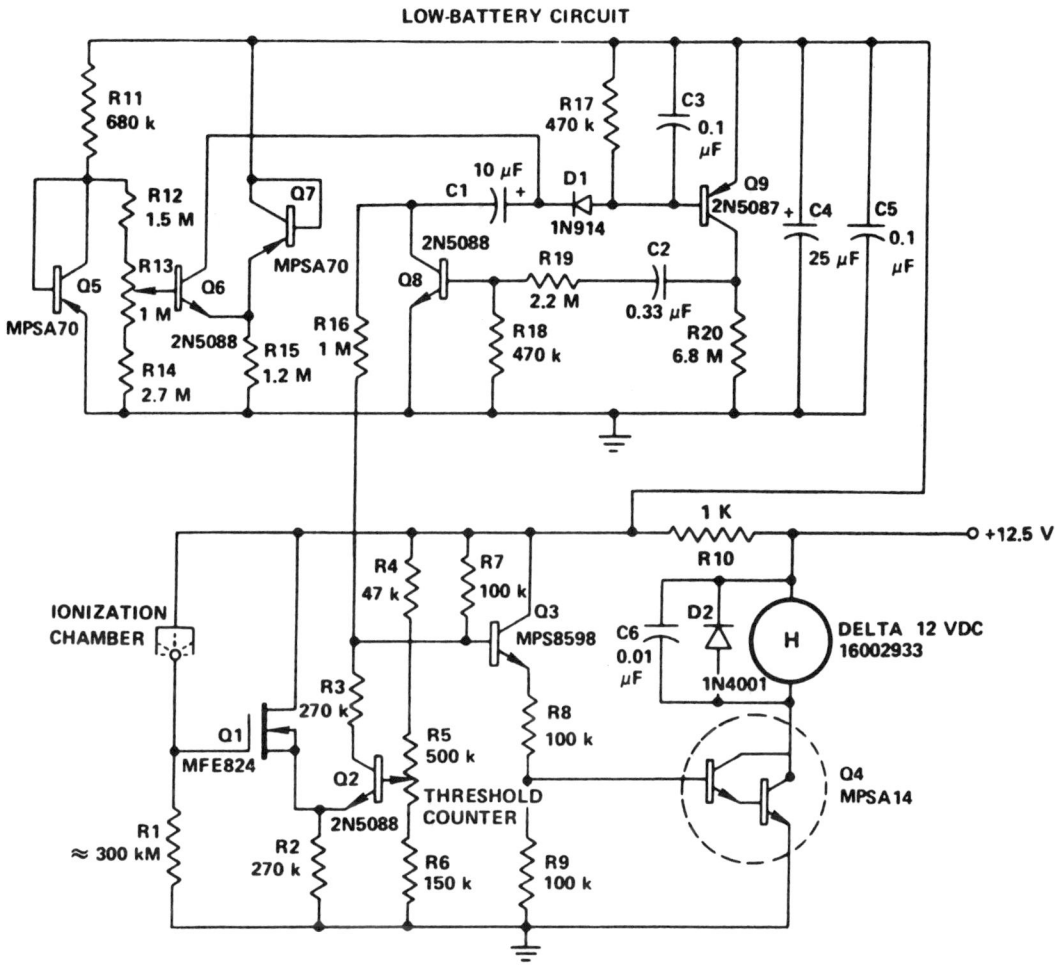

MOTOROLA

Fig. 8-14

If the smoke-alarm signal must be continuous, rather than pulsating, then the slightly less expensive, all-discrete transistor version of the MC14572 can be used.

PHOTOELECTRIC SMOKE DETECTOR (NONLATCHING)

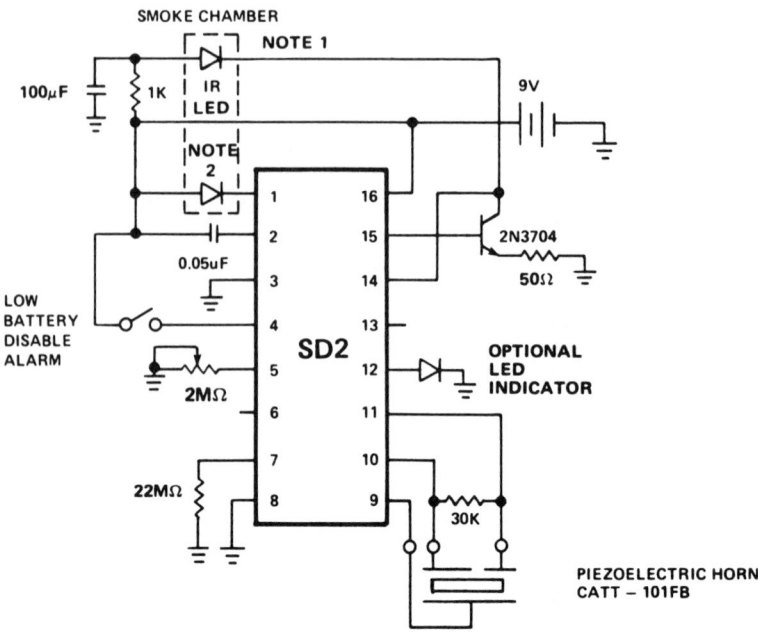

Notes: 1. IR Diode RCA Type SG 1010A or Spectronics Type SE 5455-4
Clairex Type CLED-1
2. IR Photo detectors Vactec VTS4085

SUPERTEX

Fig. 8-15

The LED predriver output pulses an external transistor, which switches on the infrared light-emitting diode at a very low duty cycle. The desired IR LED pulse period is determined by the value of the external timing resistor. The smoke sensitivity is adjustable through a trimmer resistor, which varies the IR LED pulse width. The light-sensing element is a silicon photovoltaic cell, which is held at near zero bias to minimize leakage currents. The circuit can detect signals as low as 1 mV and generate an alarm. The IR LED pulse repetition rate increases when smoke is detected.

1.9-V BATTERY-OPERATED IONIZATION-TYPE SMOKE DETECTOR

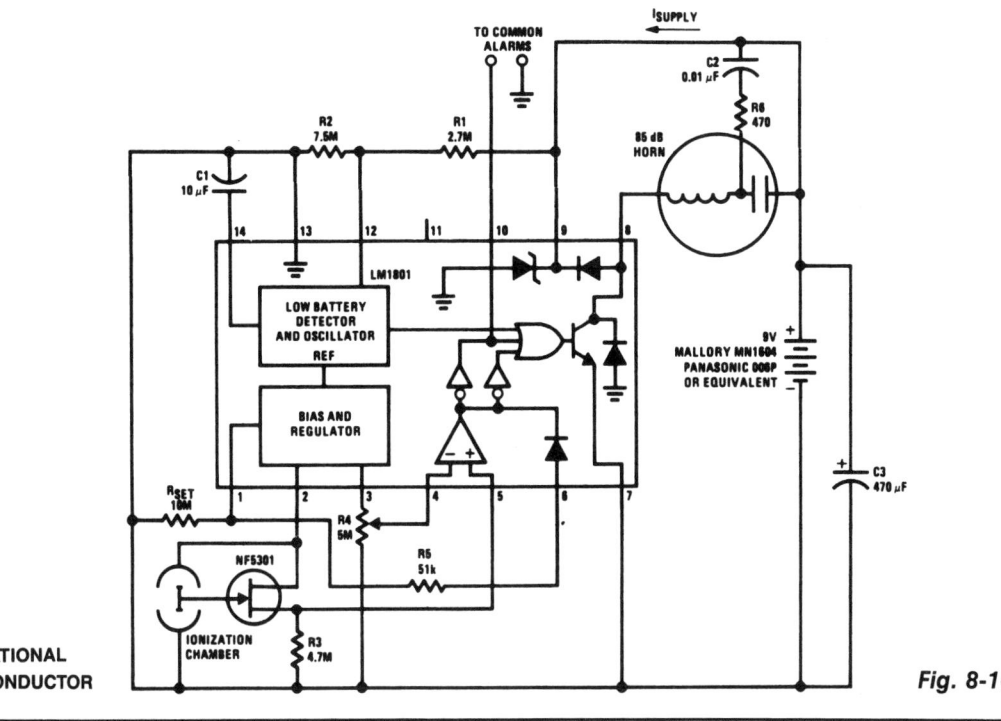

NATIONAL
SEMICONDUCTOR

Fig. 8-16

LINE-OPERATED PHOTOELECTRIC SMOKE ALARM

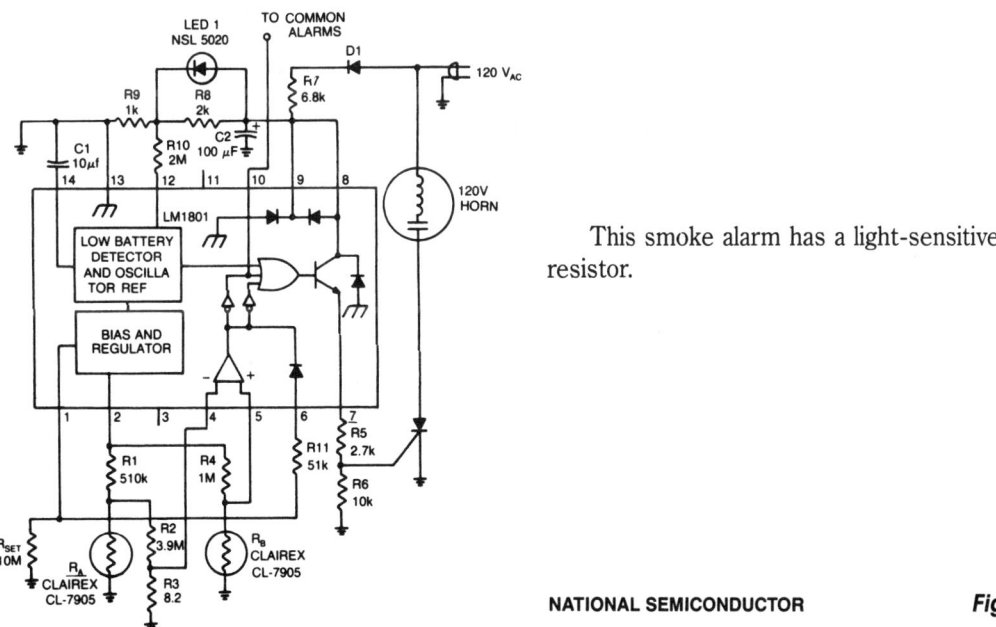

This smoke alarm has a light-sensitive resistor.

NATIONAL SEMICONDUCTOR

Fig. 8-17

SMOKE DETECTOR III

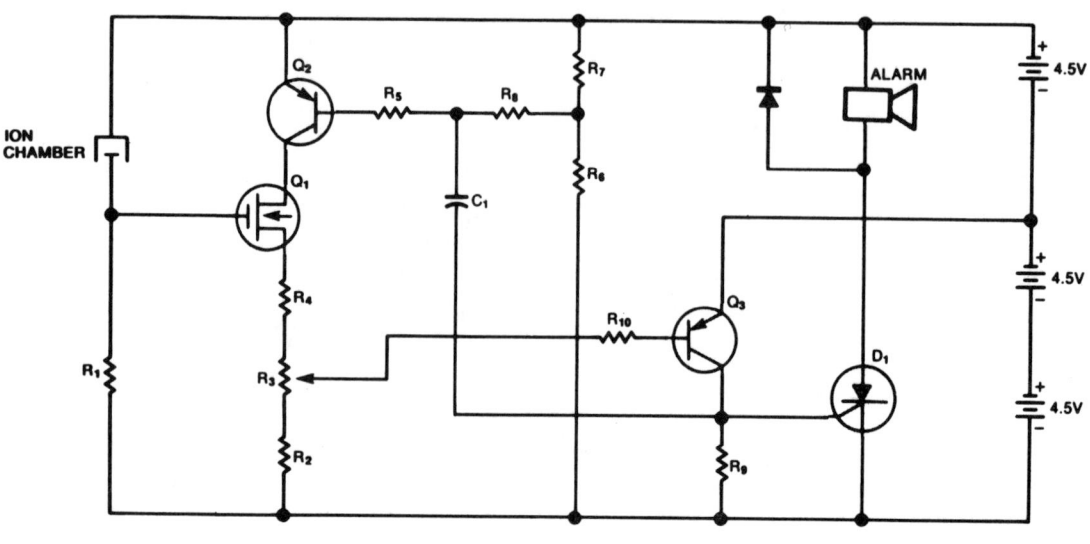

EDN

Fig. 8-18

This circuit comes from U.S. Patent 3,778,800, granted to BRK Electronics in Aurora, IL. The circuit provides a smoke detector with an alarm for both smoke and low batteries. The R6/R7 voltage divider monitors the battery and will turn Q2 and Q1 off when the battery voltage falls too low. The smoke-detector chamber will also cut Q1 off when it senses smoke. Q1 via Q3, triggers SCR D1 and sounds the alarm. Capacitor C1 provides feedback that causes the alarm to sound intermittently. The smoke detector and low-battery circuits sound the alarm at two different rates.

9

Hall-Effect Detectors

The sources of the following circuits are contained in the Sources section, which begins on page 214. The figure number in the box of each circuit correlates to the source entry in the Sources section.

Current Monitor
Security Door-Ajar Alarm
Hall-Effect Switches

Angle-of-Rotation Detector
Door-Open Alarm
Hall-Effect Compass

CURRENT MONITOR

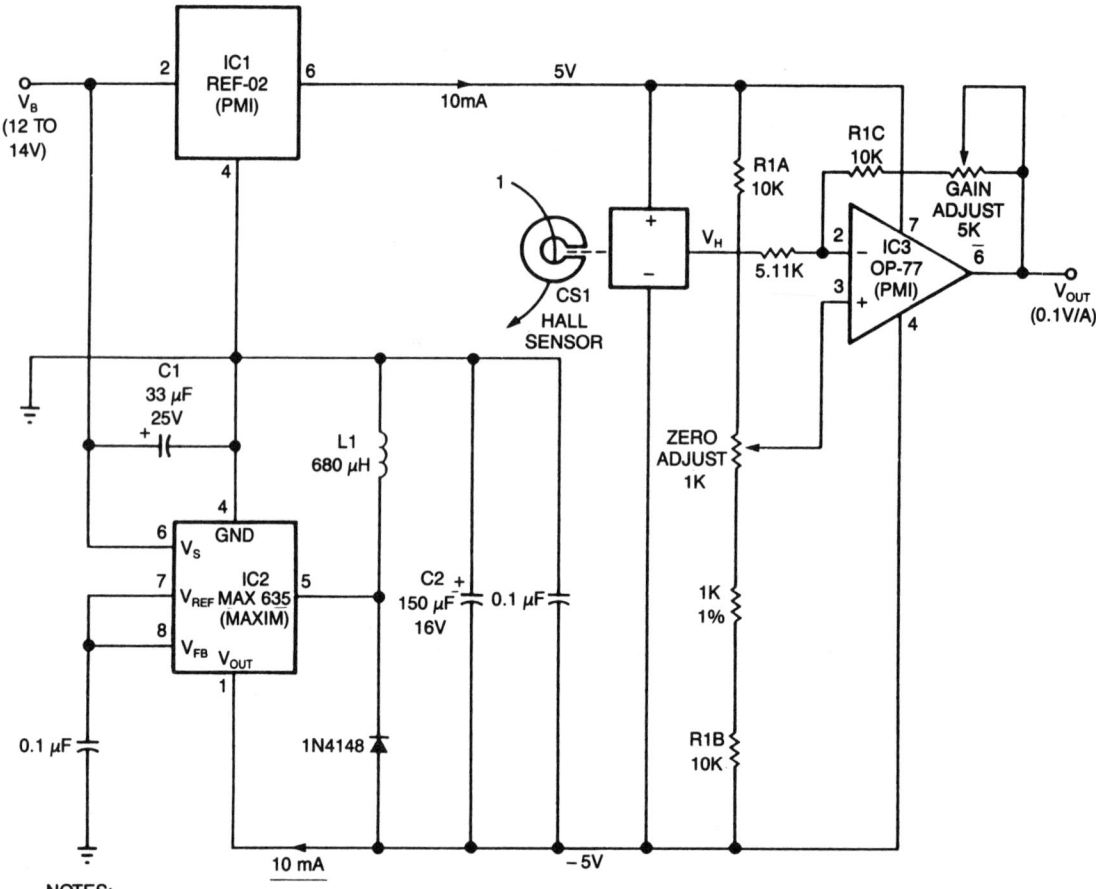

NOTES:
1. C1 AND C1 ARE 199D TANTALEX CAPACITORS FROM SPRAGUE
2. L1 IS A 6860-23 INDUCTOR FROM CADDELL-BURNS
3. R1A, R1B, AND R1C ARE PART OF A THIN-FILM RESISTOR NETWORK SUCH AS THE CADDOCK T914-10K.
4. CS1 IS A HALL-EFFECT CURRENT SENSOR (CSLA1CD) FROM MICROSWITCH.

EDN

Fig. 9-1

This circuit uses a Hall-effect sensor, consisting of an IC that resides in a small gap in a flux-collector toroid, to measure dc current in the range of 0 to 40 A. You wrap the current-carrying wire through the toroid; the Hall voltage V_H is then linearly proportional to current I. The current drain from V_B is less than 30 mA.

To monitor an automobile alternator's output current, for example, connect the car battery between the circuit's V_B terminal and ground, and wrap one turn of wire through the toroid. Or, you could wrap 10 turns—if they fit—to measure 1 A full scale. When $I = 0$ V current sensor CS_I's V_H output equals one-half of its 10 V bias voltage. Because regulators IC1 and IC2 provide a bipolar bias voltage, V_H and V_{OUT} are zero when I is zero; you can then adjust the output gain and offset to scale V_{OUT} at 1 V per 10 A.

SECURITY DOOR-AJAR ALARM

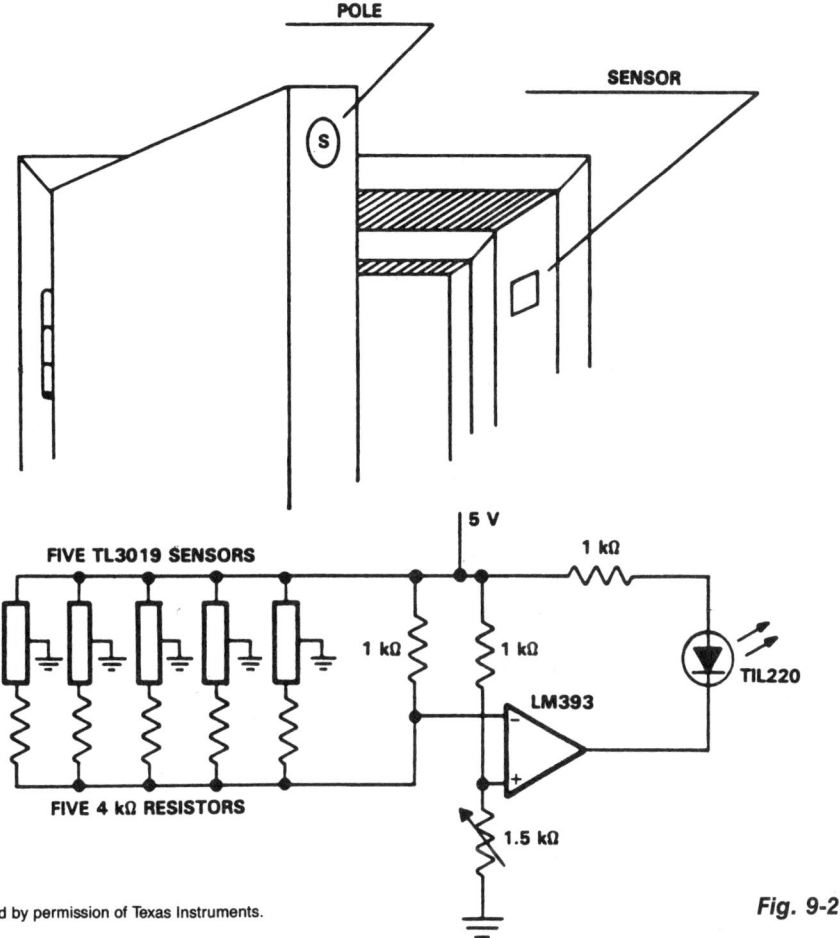

Reprinted by permission of Texas Instruments.

Fig. 9-2

In operation, the TL3019 device will activate, or become low, when a south pole of a magnet comes near the chip face of the device. The example shows five doors. Each door has a magnet embedded in its edge with the south pole facing the outer surface. At the point where the magnet is positioned with the door closed, a TL3019 sensor is placed in the door jamb. With the door closed, the Hall devices will be in a logic low state. This design has five doors and uses five TL3019 devices. Each TL3019 has a 4-kΩ resistor in series and all door sensor and resistor sets are in parallel and connected to the inverting input of an LM393 comparator. With all doors closed, the effective resistance will be about 800 Ω and produce 2.2 V at the inverting input. The noninverting input goes to a voltage divider network which sets the reference voltage. The 1.5-kΩ potentiometer is adjusted so the indicator goes out with all doors closed. This will cause 2.35 V to appear at the noninverting input of the comparator. When a door opens, the voltage at the inverting input will go to 2.5 V, which is greater than V_{REF}, and the LED will light. A large number of doors and windows can be monitored with this type of circuit. Also, it could be expanded to add an audible alarm in addition to the visual LED.

HALL-EFFECT SWITCHES

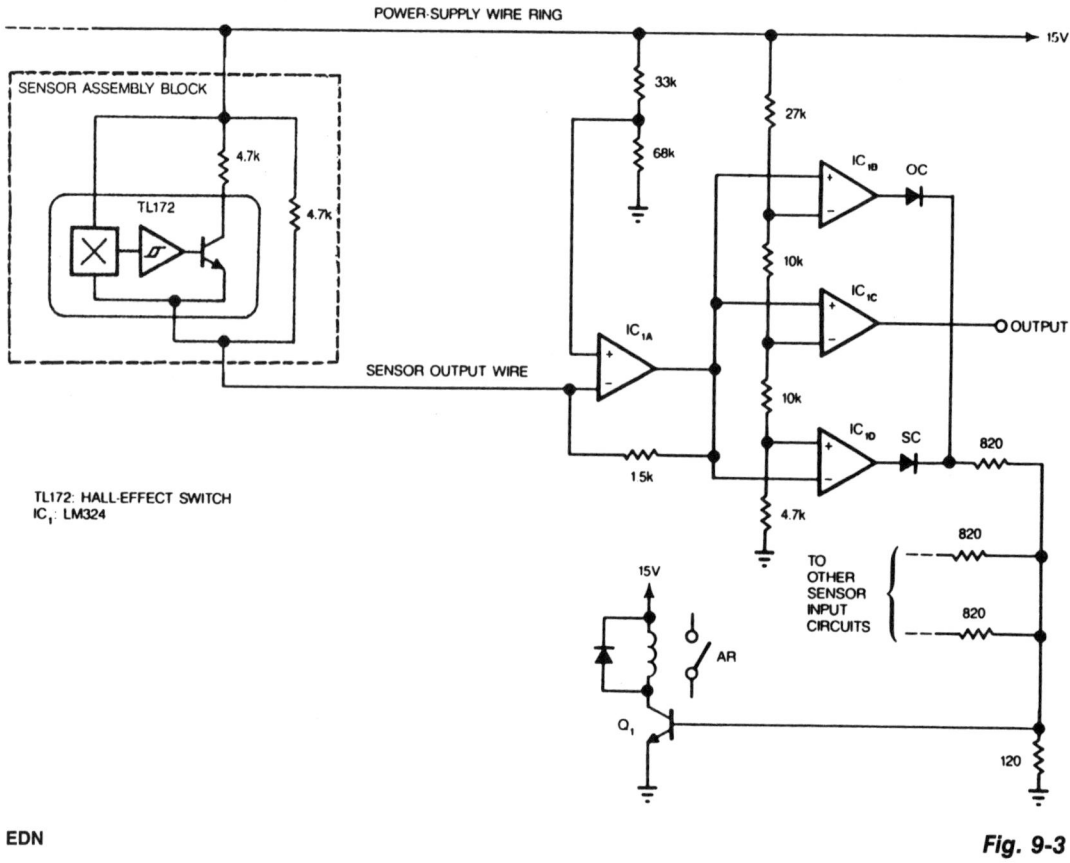

EDN

Fig. 9-3

Hall-effect switches have several advantages over mechanical and optically coupled switches. They're insensitive to environmental light and dirt, they don't bind, and they don't sustain mechanical wear. Their major drawback is that they require three wires per device. The circuit shown, however, reduces this wire count to $N+1$ wires for N devices.

Amplifier IC1A is configured as a current-to-voltage converter. It senses the sensor assembly's output current. When the Hall-effect switch is actuated, the sensor's output current increases to twice its quiescent value. Amplifier IC1B, configured as a comparator, detects this increase. The comparator's output decreases when the Hall-effect switch turns on.

The circuit also contains a fault-detection function. If any sensor output wire is open, its corresponding LED will turn on. If the power-supply line opens, several LEDs will turn on. A short circuit will also turn an LED on. Every time an LED turns on, Q1 turns on and the alarm relay is actuated.

ANGLE-OF-ROTATION DETECTOR

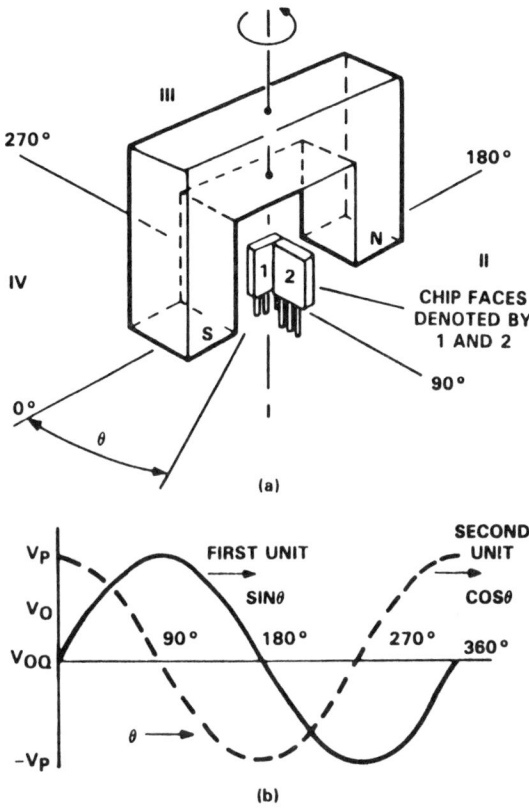

(a)

(b)

TEXAS INSTRUMENTS

Fig. 9-4

The figure shows two TL3103 linear Hall-effect devices used for detecting the angle of rotation. The TL3103s are centered in the gap of a U-shaped permanent magnet. The angle that the south pole makes with the chip face of unit #1 is defined as angle 0. Angle 0 is set to 0° when the chip face of unit #1 is perpendicular to the south pole of the magnet. As the south pole of the magnet sweeps through a 0° to 90° angle, the output of the sensor increases from 0°. Sensor unit #2 decreases from its peak value of $+V_p$ at 0° to a value V_{OQ} at 90°. So, the output of sensor unit #1 is a sine function of 0 and the output of unit #2 is a cosine function of 0 as shown. Thus, the first sensor yields the angle of rotation and the second sensor indicates the quadrant location.

DOOR-OPEN ALARM

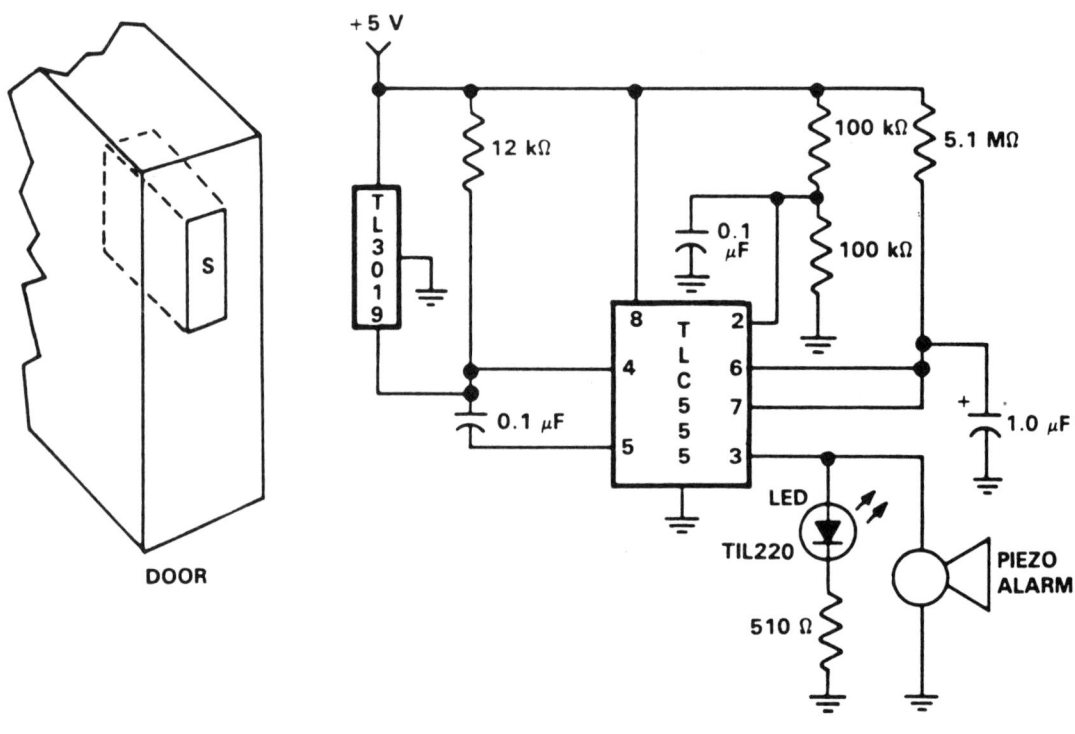

TEXAS INSTRUMENTS

Fig. 9-5

Door-open alarms are used chiefly in automotive, industrial, and appliance applications. This type of circuit can sense the opening of a refrigerator door. When the door opens, a triac could be activated to control the inside light. The figure shows a door position alarm. When the door is opened, an LED turns on and the piezo alarm sounds for approximately 5 seconds. This circuit uses a TL3019 Hall-effect device for the door sensor. This normally open switch is located in the door frame. The magnet is mounted in the door.

When the door is in the closed position, the TL3019 output goes to logic low, and remains low until the door is opened. This design consists of a TLC555 monostable timer circuit. The 1-μF capacitor and 5.1-MΩ resistor on pins 6 and 7 set the monostable RC time constant. These values allow the LED and piezo alarm to remain on about 5 seconds when triggered.

One unusual aspect of this circuit is the method of triggering. Usually, a 555 timer circuit is triggered by taking the trigger, pin 2, low, which produces a high at the output, pin 3. In this configuration, with the door in the closed position, the TL3019 output is held low. The trigger, pin 2, is connected to ½ the supply voltage V_{CC}. When the door opens, a positive high pulse is applied to control pin 5 through a 0.1-μF capacitor and also to reset pin 4. This starts the timing cycle. Both the piezo alarm and the LED visual indicator are activated.

HALL-EFFECT COMPASS

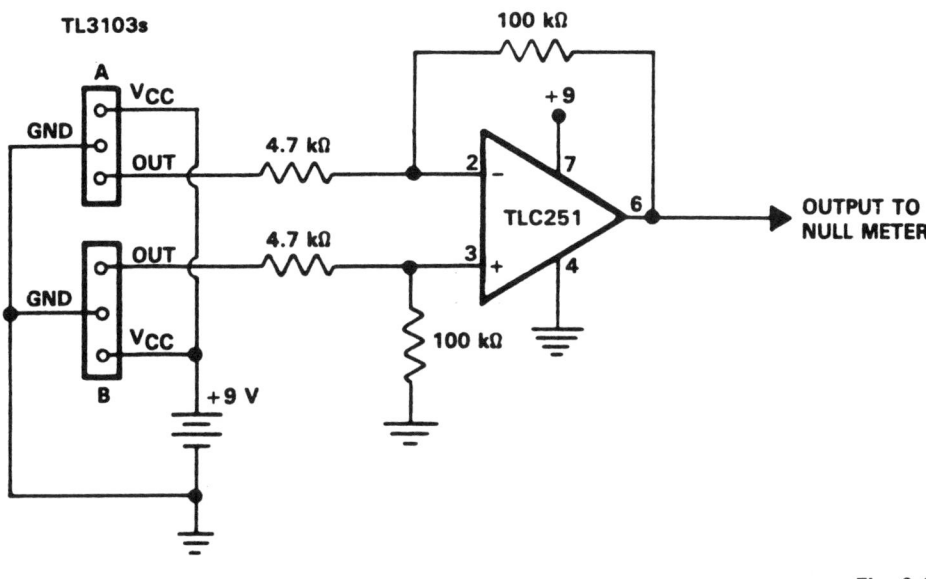

TEXAS INSTRUMENTS

Fig. 9-6

The TL3103 linear hall-effect device can be used as a compass. By definition, the north pole of a magnet is the pole that is attracted by the magnetic north pole of the earth. The north pole of a magnet repels the north-seeking pole of a compass. By convention, lines of flux emanate from the north pole of a magnet and enter the south pole. The circuit of the compass is shown. By using two TL3103 devices instead of one, we achieve twice the sensitivity.

With each device facing the opposite direction, device A would have a position output while the output of device B would be negative, with respect to the zero-magnetic field level. This gives a differential signal to apply to the TLC251 op amp. The op amp is connected as a difference amplifier with a gain of 20. Its output is applied to a null meter or a bridge-balance indicator circuit.

10

Ice-Formation Detectors

The sources of the following circuits are contained in the Sources section, which begins on page 214. The figure number in the box of each circuit correlates to the source entry in the Sources section.

Ice-Formation Alarm
Road Ice Alarm
Ice Warning and Lights Reminder

ICE-FORMATION ALARM

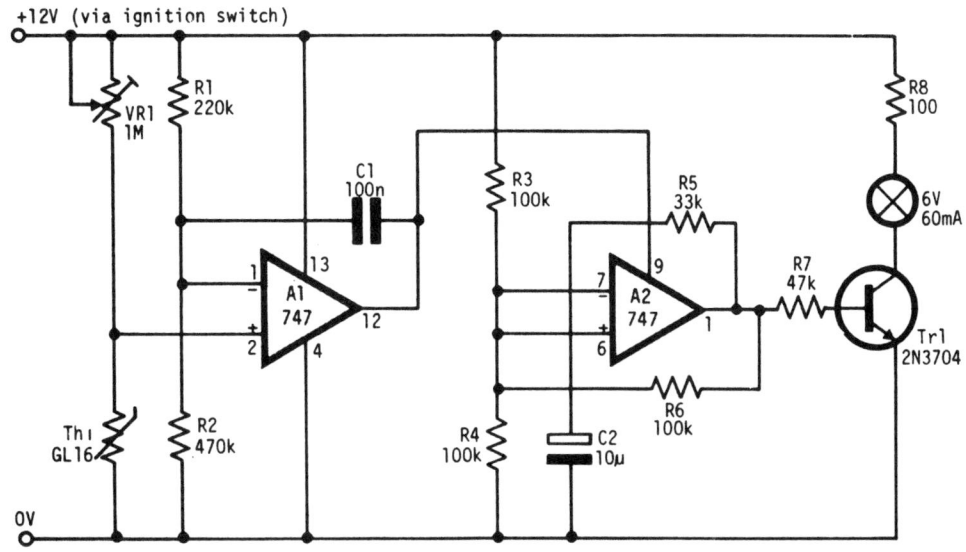

ELECTRONIC ENGINEERING

Fig. 10-1

The circuit warns car drivers when the air temperature close to the ground approaches 0°C, thereby indicating possible formation of ice on the road surface. Op amp A1 is wired as a voltage-level sensor. Op amp A2 is wired as an astable multivibrator, which, by means of current buffer Tr1, flashes a filament lamp at about 1 Hz. As air temperature falls, a point is reached when the voltage at pin 2 just rises above the voltage at pin 1. The output of A1 is immediately driven into positive saturation, because it is operated open loop. This positive output voltage powers A2 through its V+ connection on pin 9, starting the oscillator. The thermistor is a glass-bead type with a resistance of about 20 MΩ at 20°C. VR1 is adjusted so that the lamp starts flashing when the air temperature is 1 to 2°C.

ROAD ICE ALARM

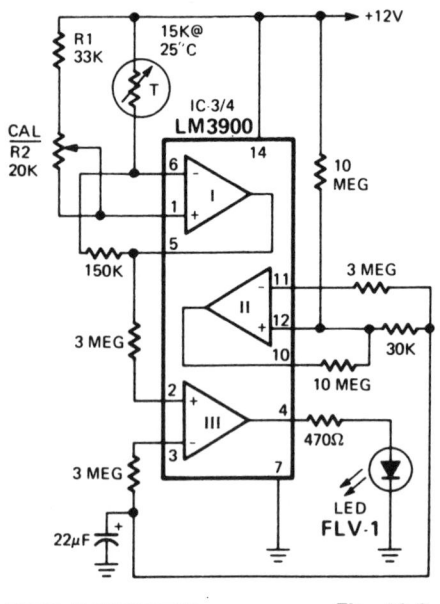

The circuit uses a thermistor and three sections of an LM3900 quad op amp IC. When the temperature drops to 36°F the LED indicator flashes about once each second. The flashing rate increases as temperature drops to 32°F when the LED remains on. Amplifier I compares the thermistor's resistance to the resistance of the standard network connected to its noninverting input. Its output—fed to the noninverting input of op amp III—varies with temperature. Op amp II is a free-running multivibrator feeding a pulse signal of about 1 Hz to the inverting input of op amp III. This amplifier compares the outputs of op amps I and II and turns on the LED when the multivibrator's output level drops below op amp I. The monitor is calibrated by placing the thermistor in a mixture of crushed ice and water and adjusting the 20-kΩ pot so that the LED stays on.

RADIO-ELECTRONICS   *Fig. 10-2*

ICE WARNING AND LIGHTS REMINDER

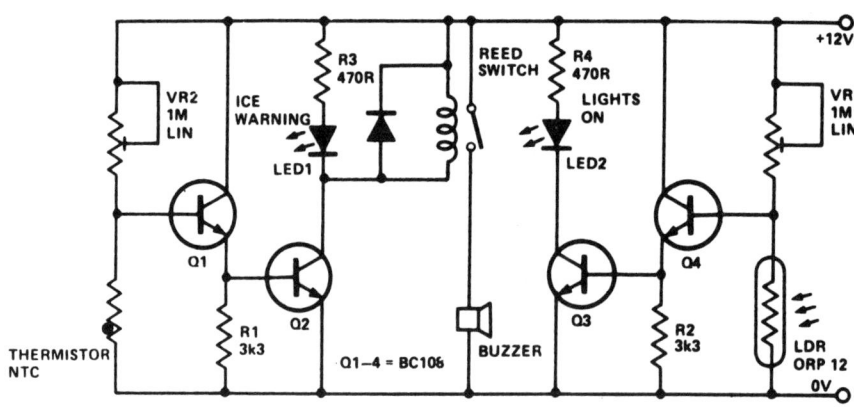

ELECTRONICS TODAY INTERNATIONAL   *Fig. 10-3*

This device will tell a driver if his lights should be on and will warn him if the outside temperature is nearing zero by lighting a LED and sounding a buzzer. VR1 adjusts sensitivity for temperature, VR2 for light. Both thermistor and LDR should be well protected. Most high-gain npn transistors will work.

11

Level Detectors

The sources of the following circuits are contained in the Sources section, which begins on page 214. The figure number in the box of each circuit correlates to the source entry in the Sources section.

Low-Voltage Detector
Level Detector with Hysteresis (Positive
    Feedback)
Precision Threshold Detector
Voltage-Level Detector

LOW-VOLTAGE DETECTOR

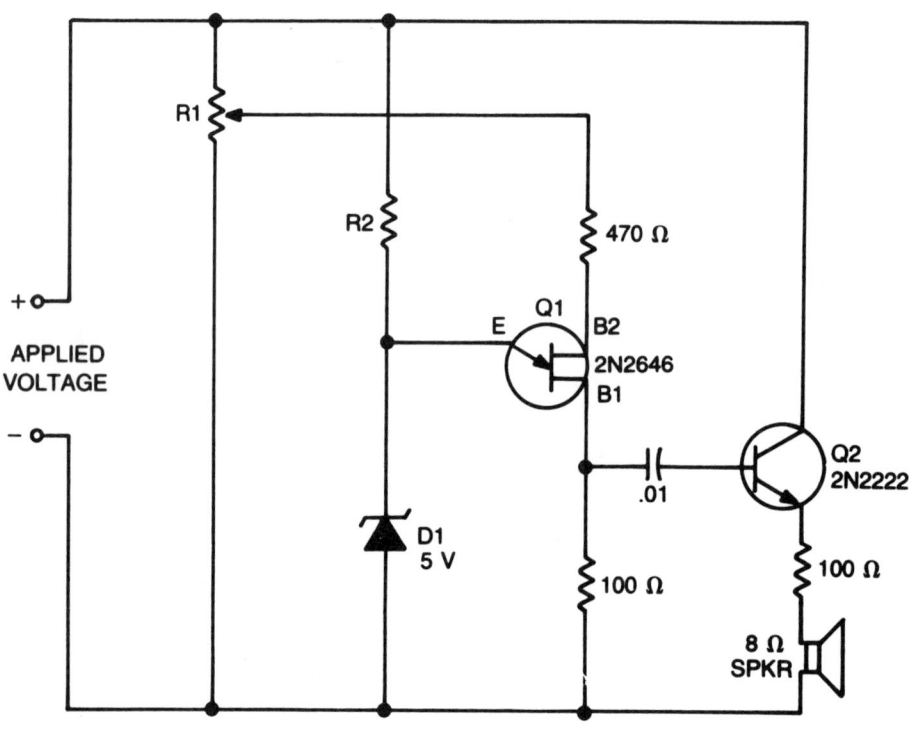

RADIO-ELECTRONICS

Fig. 11-1

The values of R_1, R_2, and D1 are selected for the voltage applied. Using a 12-volt battery, $R_1 = 10$ kΩ, $R_2 = 5.6$ kΩ and D1 is a 5-V zener diode, or a string of forward-biased silicon rectifiers equaling about 5 V. Transistor Q1 is a general-purpose UJT (unijunction transistor), and Q2 is any small-signal or switching npn transistor. When the detector is connected across the battery terminals, it draws little current and does not interfere with other devices powered by the battery. If voltage drops below the trip voltage selected with the R1 setting, the speaker beeps a warning. The frequency of the beeps is determined by the amount of undervoltage. If other voltages are being monitored, select R1 so that it draws only 1 mA or 2 mA. Zener diode D1 is about one-half of the desired trip voltage, and R2 is selected to bias it at about 1 mA.

LEVEL DETECTOR WITH
HYSTERESIS (POSITIVE FEEDBACK)

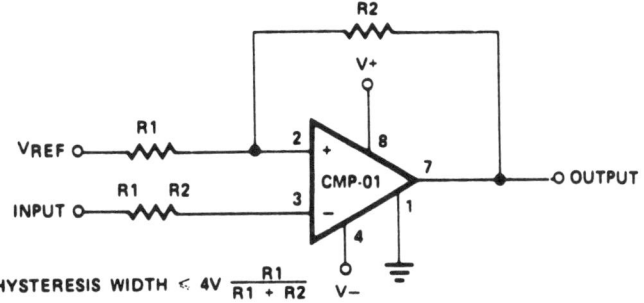

$$\text{HYSTERESIS WIDTH} \leqslant 4V \frac{R1}{R1 + R2}$$

PRECISION MONOLITHICS

Fig. 11-2

PRECISION THRESHOLD DETECTOR

*INPUT RESISTORS NECESSARY IF DIFFERENTIAL
INPUT VOLTAGE EXCEEDS ±1V.

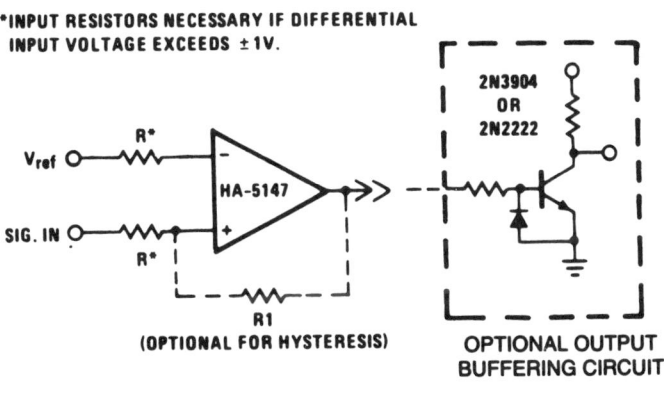

R1
(OPTIONAL FOR HYSTERESIS)

OPTIONAL OUTPUT
BUFFERING CIRCUIT

HARRIS

Fig. 11-3

This circuit requires low noise, low and stable offset voltages, high open-loop gain, and high speed. These requirements are met by the HA-5147. The standard variations of this circuit can easily be implemented using the HA-5147. For example, hysteresis can be generated by adding R1 to provide small amounts of positive feedback. The circuit becomes a pulse-width modulator if V_{ref} and the input signal are left to vary. Although the output-drive capability of this device is excellent, the optional buffering circuit can be used to drive heavier loads, preventing loading effects on the amplifier.

VOLTAGE-LEVEL DETECTOR

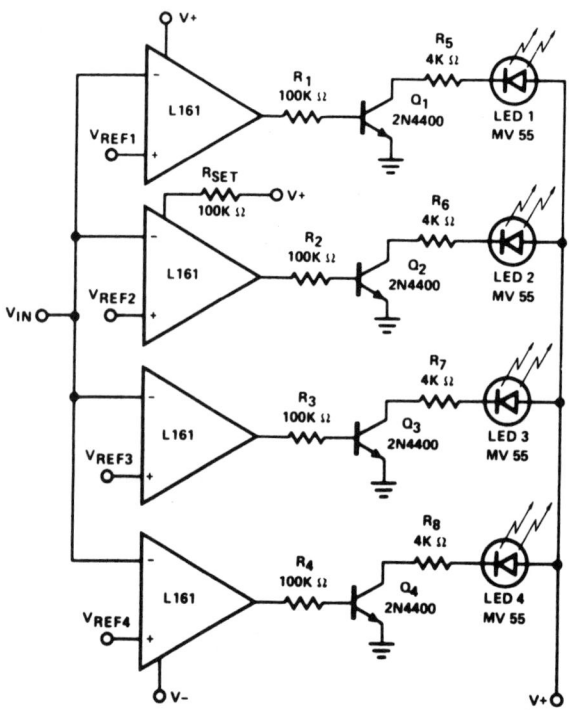

SILICONIX

Fig. 11-4

12

Lie Detector

The source of the following circuit is contained in the Sources section, which begins on page 214. The figure number contained in the box of the circuit correlates to the source entry in the Sources section.

Lie Detector

LIE DETECTOR

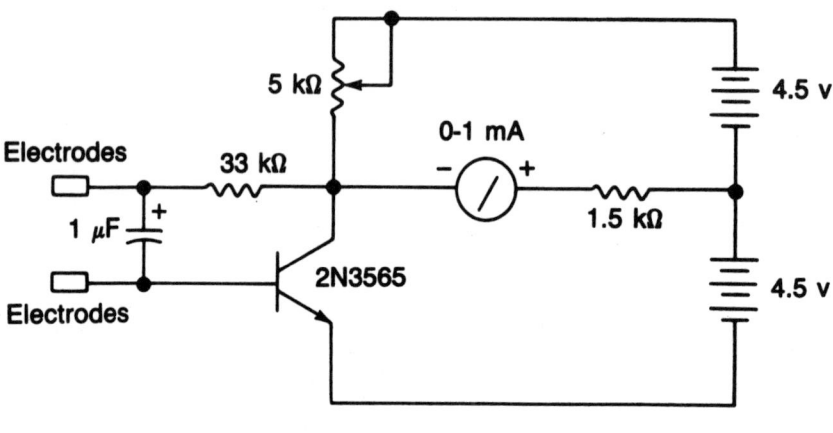

WILLIAM SHEETS

Fig. 12-1

The two probes shown are held in the hands and the skin resistance applies bias to the transistor. The 5-kΩ pot is set for zero deflection on the meter. When the "subject" is embarrassed or lies, sweating of the hands occurs, increasing the bias to the transistor and upsetting the bridge balance.

13

Liquid Detectors

The sources of the following circuits are contained in the Sources section, which begins on page 214. The figure number in the box of each circuit correlates to the source entry in the Sources section.

Flood Alarm
Liquid-Level Detector
Liquid-Level Control
Latching Liquid-Level Detector
Water-Level Sensor and Control
Single-Chip Pump Controller
Liquid-Level Checker
Fluid-Level Control
Flood Alarm or Temperature Monitor

Dual Liquid-Level Detector
Level Sensor for Cryogenic Fluids
Fluid-Level Controller
High-Level Warning Device
Liquid-Level Monitor
Water-Level Indicator
Water-Level Alarm
Low-Level Warning with Audio Output

FLOOD ALARM

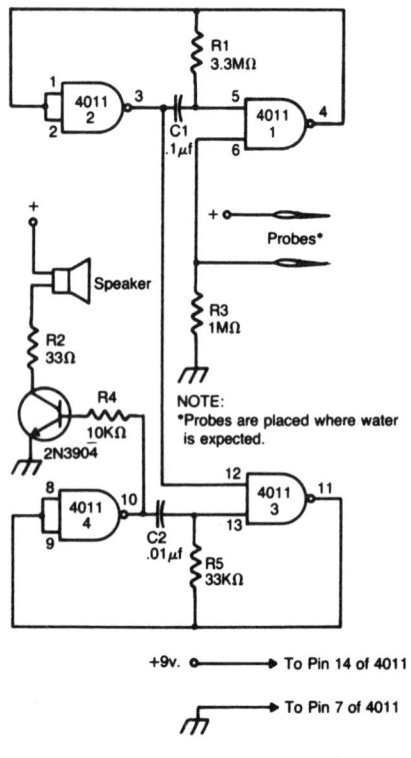

The alarm is built around two audio oscillators, each using two NAND gates. The detection oscillator is gated on by a pair of remote probes. One of the probes is connected to the battery supply, the other to the input of one of the gates. When water flows between the probes, the detection oscillator is gated on. The alarm oscillator is gated on by the output of the detection oscillator. The values given produce an audio tone of about 3000 Hz. The detection oscillator gates this audio tone at a rate of about 3 Hz. The result is a unique pulsating note. Use any 8-Ω speaker to sound the alarm. The 2N3904 can be replaced by any similar npn transistor. The circuit will work from any 6 to 12-V supply.

MODERN ELECTRONICS   *Fig. 13-1*

LIQUID-LEVEL DETECTOR

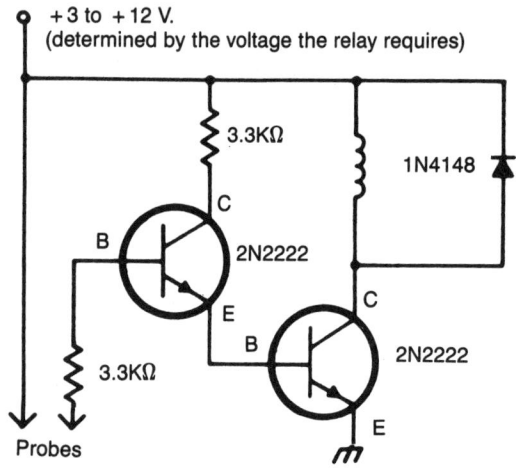

When liquid level reaches both probes, alarm is turned on. When water level recedes, the alarm goes off.

MODERN ELECTRONICS   *Fig. 13-2*

LIQUID-LEVEL CONTROL

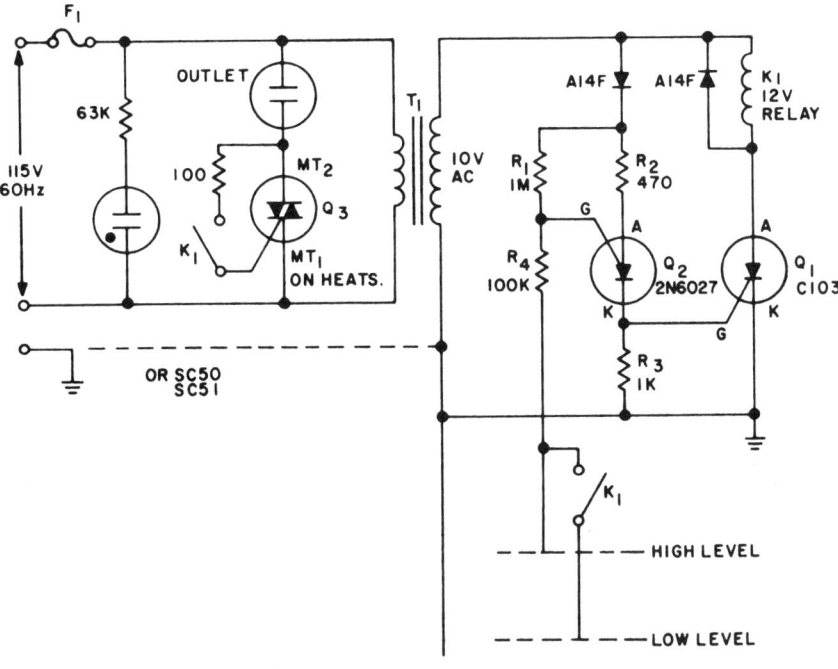

Fig. 13-3

Use this circuit to keep the fluid level of a liquid between two fixed points. Two modes, for filling or emptying are possible by simply reversing the contact connections of K1. The loads can be either electric motors or solenoid-operated valves, operating from ac power. Liquid-level detection is accomplished by two metal probes, one measuring the high level and the other the low level. An inversion of the logic (keeping the container filled) can be accomplished by replacing the normally open contact on the gate of Q3 with a normally closed contact.

LATCHING LIQUID-LEVEL DETECTOR

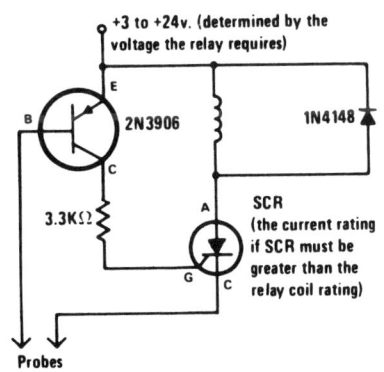

Fig.13-4

Alarm is actuated when the liquid level is above the probes and remains activated, even if the level drops below the probes. This latching action lets you know that the preset level has been reached or exceeded sometime in the past.

WATER-LEVEL SENSOR AND CONTROL

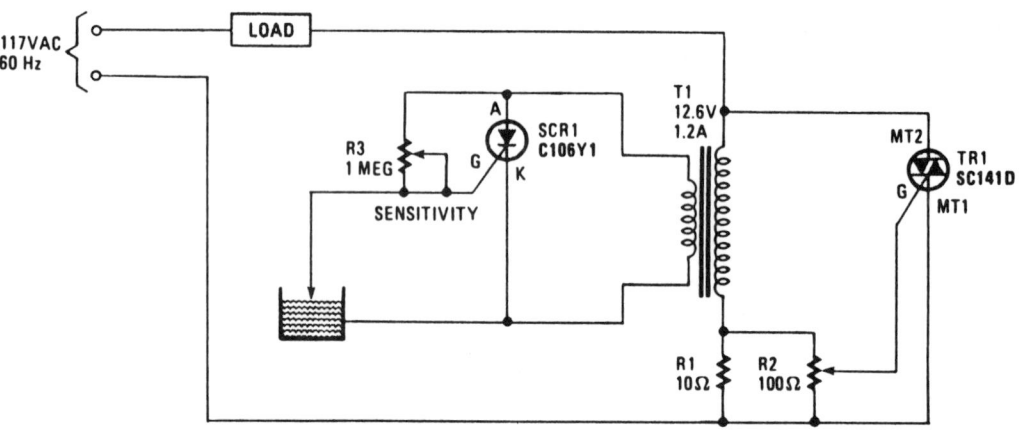

Fig. 13-5

The operation of the circuit is based on the difference in the primary impedance of a transformer when its secondary is loaded and when it is an open circuit. The impedance of the primary of T1 and resistor R1 are in series with the load. The triac's gate-control voltage is developed across parallel resistors R1 and R2. When the water level is low, the probe is out of the water and SCR1 is triggered on. It conducts and imposes a heavy load on transformer T1's secondary winding. That load is reflected back into the primary, gating triac TR1 on, which energizes the load. If the load is an electric value in the water-supply line, it will open and remain open until the water rises and touches the probe, which shorts SCR1's gate and cathode, thereby turning off the SCR1, which effectively open-circuits the secondary. The open-circuit condition—when reflected back to the primary winding—removes the triac's trigger signal, thereby turning the water off. The load might range from a water valve to a relay that controls a pump supplying water for irrigation, or a solenoid valve controlling the water level in a garden-lily pond.

SINGLE-CHIP PUMP CONTROLLER

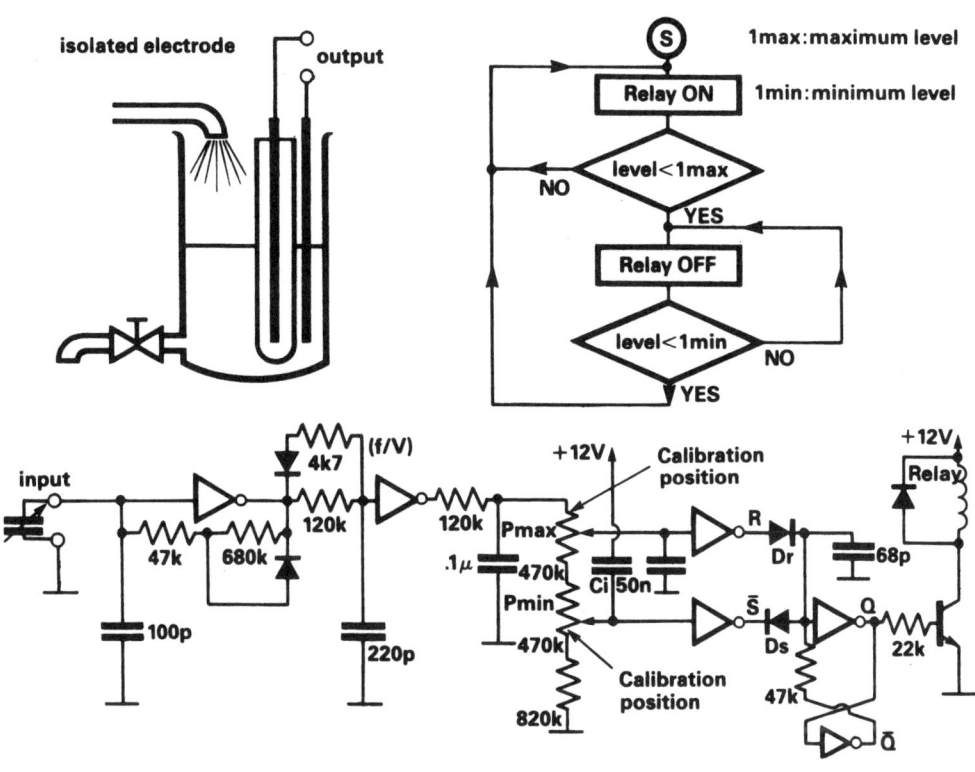

ELECTRONIC ENGINEERING

Fig. 13-6

This circuit controls the level of a tank using a bang-bang controlled electrical pump. The actual level of liquid is measured by a capacitive level meter. The first inverter performs as a capacitance-to-frequency converter. It is a Schmitt oscillator and its frequency output decreases as the capacitance increases. The second inverter is a monostable which performs as a frequency-to-voltage converter (f/V). Its output is applied to the maximum- and minimum-level comparator inputs. Maximum and minimum liquid levels can be set by the potentiometers. The maximum level (1 max.) can be preset between the limits: 65 pF less than C (1 max.) less than 120 pF. The minimum level is presetable and the limits are: 0 less than C (1 min.) less than 25 pF.

LIQUID-LEVEL CHECKER

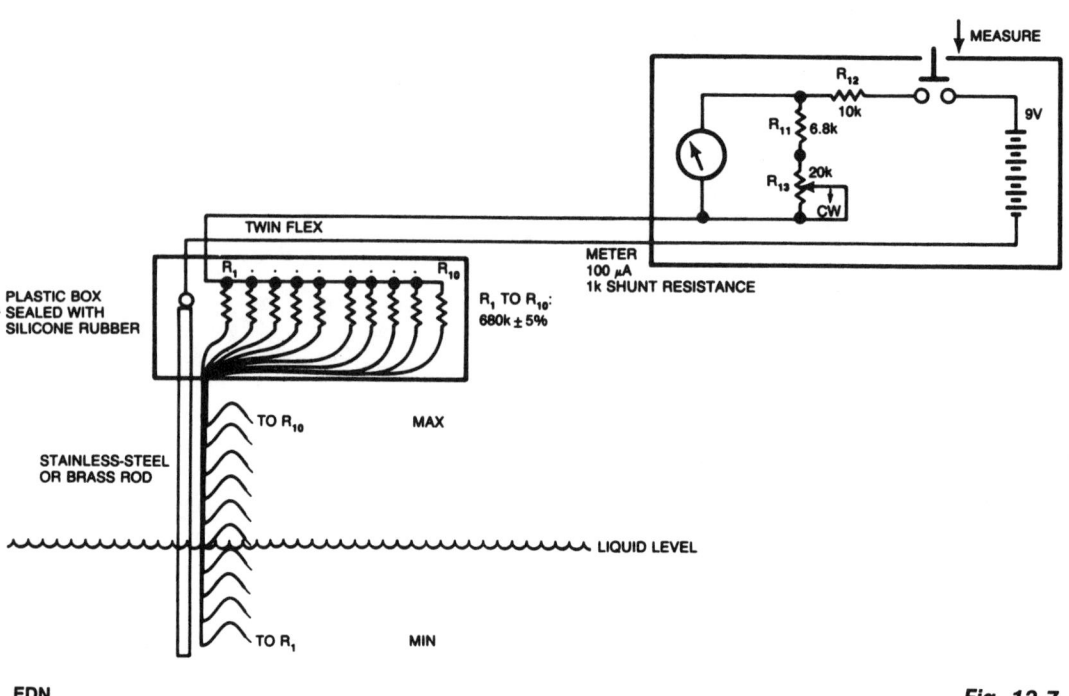

EDN

Fig. 13-7

Although many circuits use the varying-capacitance method for checking liquid levels, this simple resistive circuit is much easier to construct. Even a tank of a liquid, such as water, has sufficient conductive salts in solution for this method to work. The probe uses a metal rod that supports 10 insulated wires, which have stripped ends pointing down. As the level of liquid rises, resistors R1 through R10 are successively brought into circuit, each drawing an extra 10 μA through the meter. Shunt resistors R11 and R13 calibrate the meter for a full-scale reading when the tank is full. Resistor R12 limits the current through the meter. If the tank isn't rectangular (i.e., if the volume of the liquid it contains isn't directly proportional to the liquid's depth) space the resistors accordingly or use a nonlinear progression of resistor values and retain constant resistor spacing.

FLUID-LEVEL CONTROL

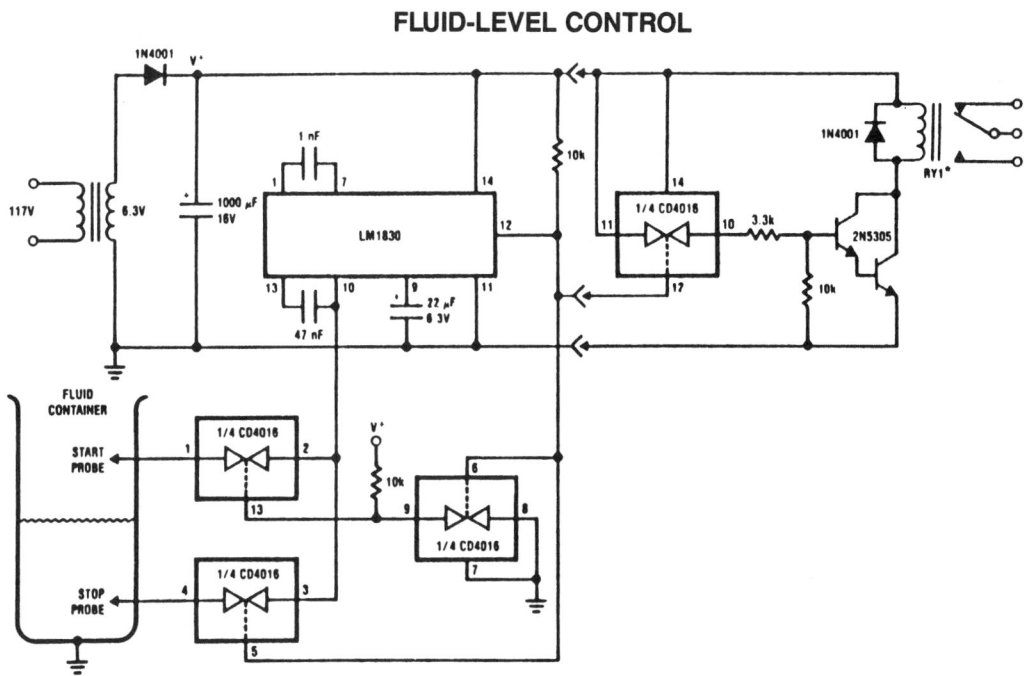

"Emptying" Processes are Controlled with this Circuit

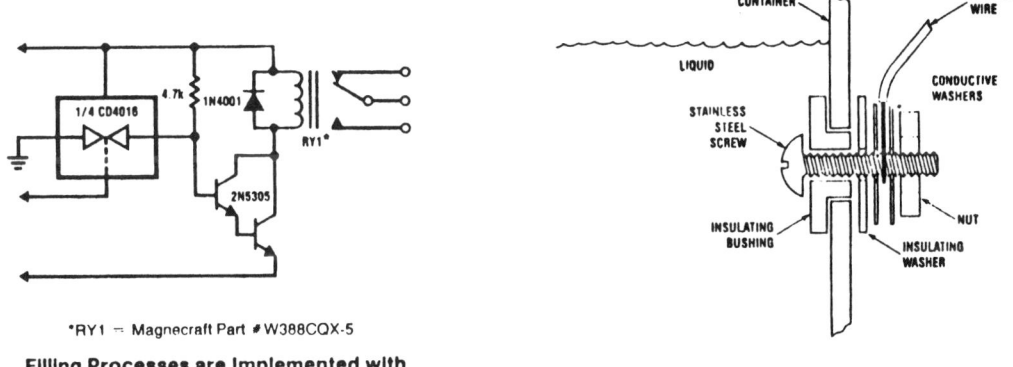

*RY1 = Magnecraft Part # W388CQX-5

Filling Processes are Implemented with this Output Circuit and Relabeled Probes

A sealing compound applied externally protects hook-up wire and prevents leaks.

NATIONAL SEMICONDUCTOR

Fig. 13-8

This circuit is designed to detect the presence or absence of aqueous fluids. An ac signal generated on-chip is passed through two probes within the fluid. A detector determines the presence of the fluid by using the probes in a voltage-divider circuit and measuring the signal level across the probes. An ac signal is used to prevent plating or dissolving of the probes as occurs when a dc signal is used. A pin is available for connecting an external resistance in cases where the fluid impedance is not compatible with the internal 13-kΩ divider resistance.

FLOOD ALARM OR TEMPERATURE MONITOR

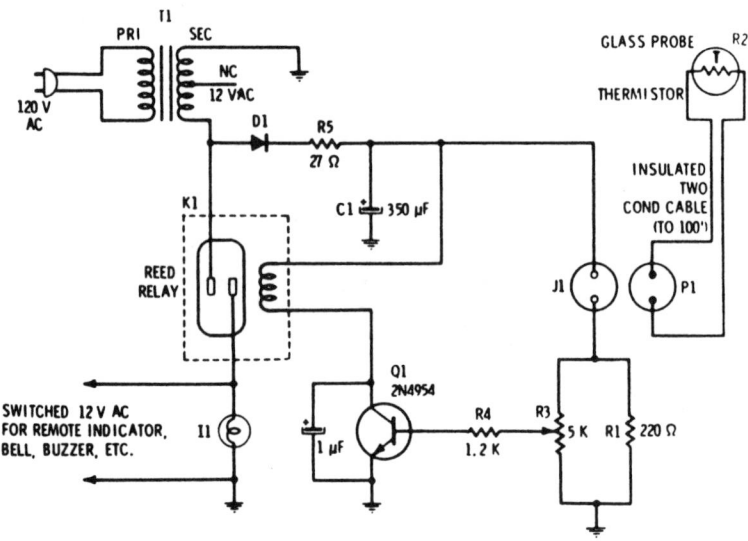

TAB BOOKS

Fig. 13-9

Filtered 15 Vdc is applied to a series circuit consisting of thermistor R2 and parallel combination of resistors R1 and R3. Transistor Q1 acts as a switch whose state is determined by the setting of potentiometer R3, which is first set so just enough current flows into the base to switch on when the thermistor is in contact with air. When the resistance of the thermistor decreases, the voltage at the base of Q1 rises. When the base current reaches the preset level, the transistor conducts and passes current through the reed relay coil, closing the reed relay contacts. The current at the base of transistor Q1 is determined by the environment into which the thermistor is inserted.

DUAL LIQUID-LEVEL DETECTOR

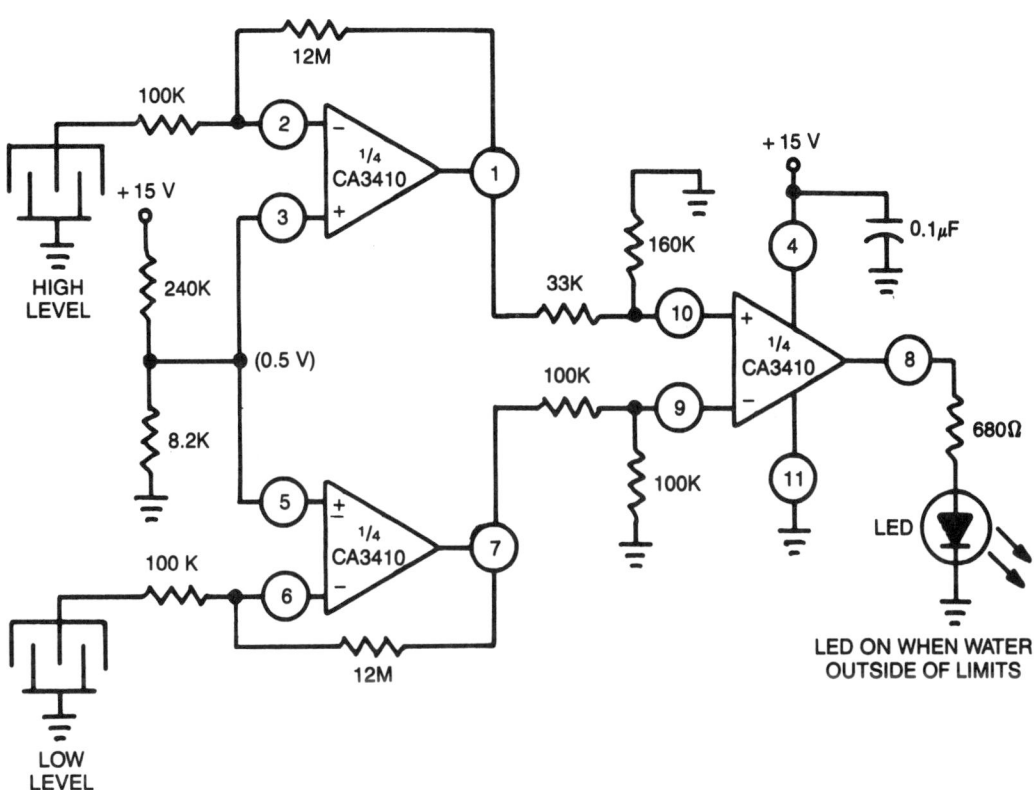

Fig. 13-10

Uses CA3410 quad BiMOS op amp to sense small currents. Because the op amp's input current is low, a current of only 1 μA passing through the sensor will change the converter's output by as much as 10 to 12 V.

LEVEL SENSOR FOR CRYOGENIC FLUIDS

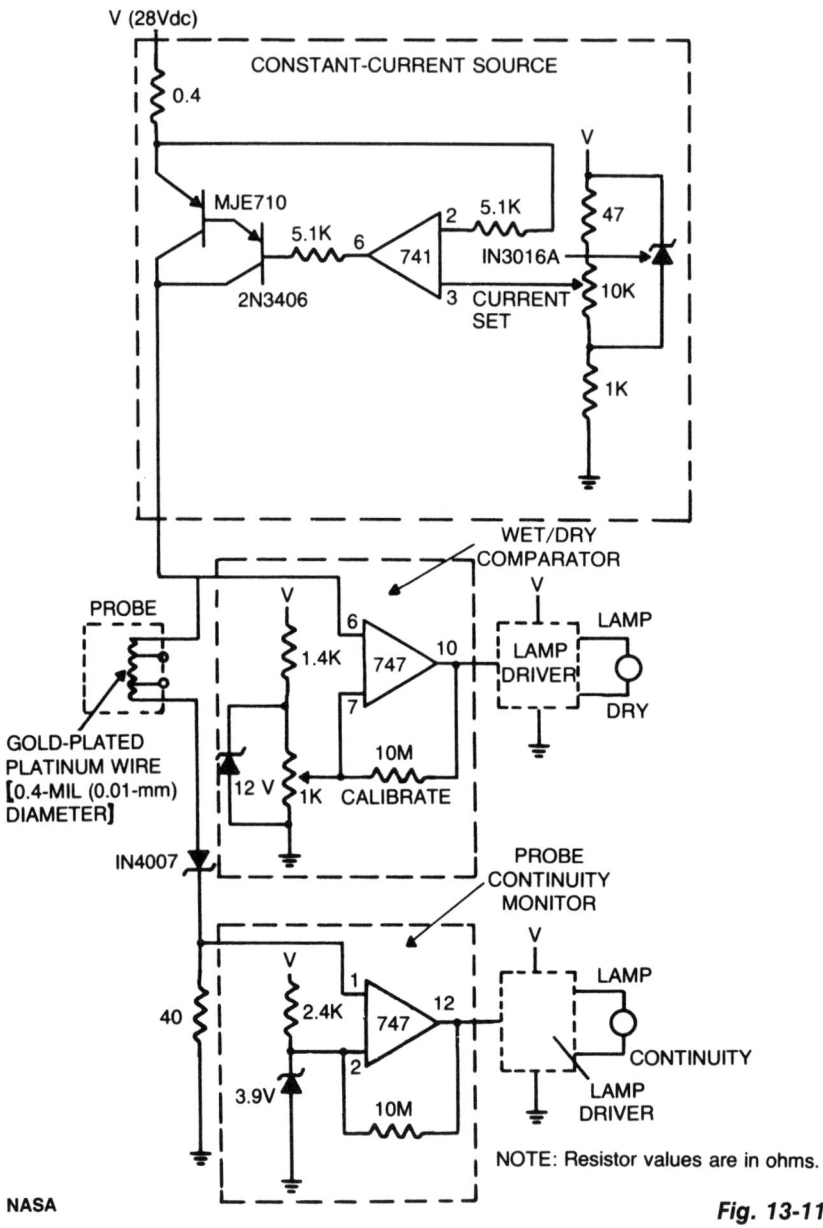

NASA

Fig. 13-11

The sensor circuit is adaptable to different liquids and sensors. The constant-current source drives current through the sensing probe and a fixed resistor. The voltage-comparator circuits interpret the voltage drops to tell whether the probe is immersed in liquid and whether current is in the probe.

FLUID-LEVEL CONTROLLER

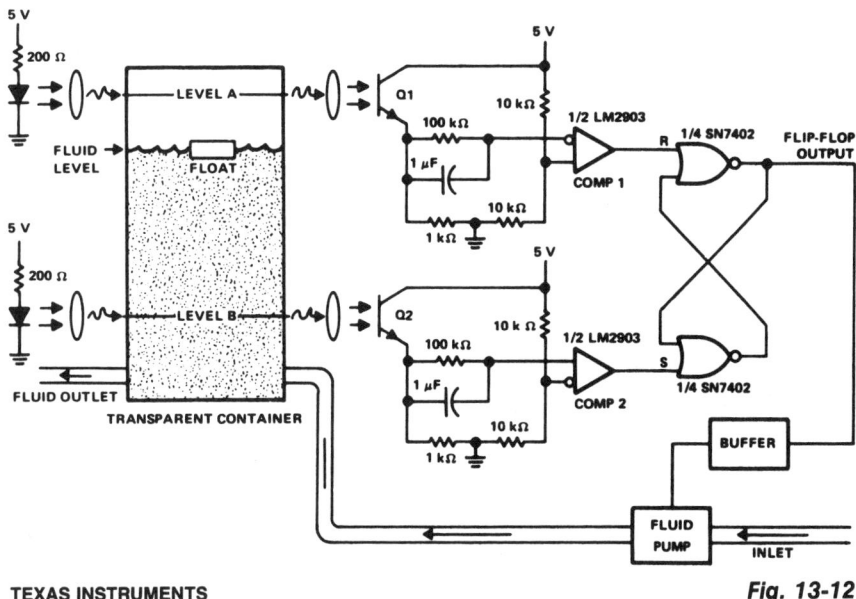

TEXAS INSTRUMENTS

Fig. 13-12

This circuit can be used to maintain fluid between two levels. Variations on this control circuit can be made to keep something that moves within certain boundary conditions.

HIGH-LEVEL WARNING DEVICE

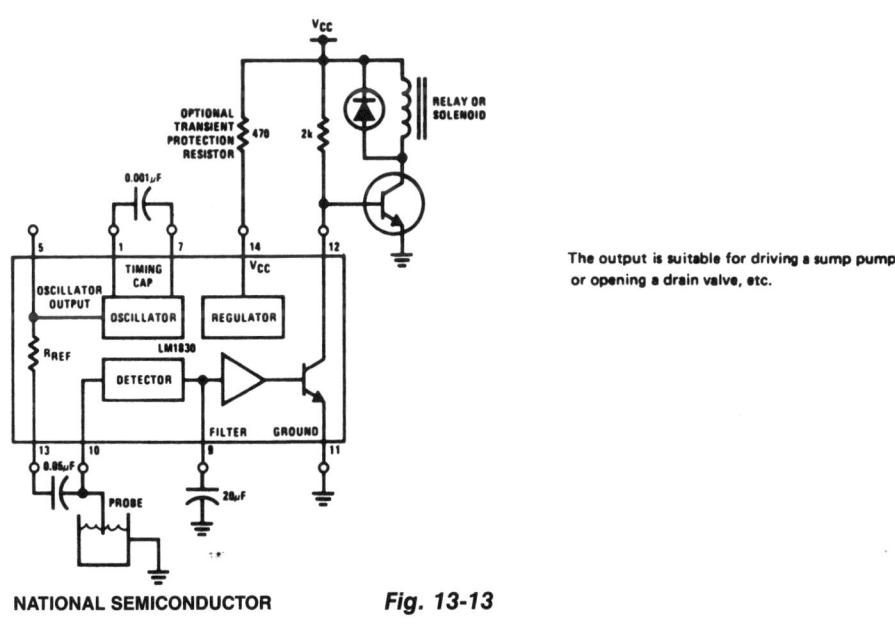

The output is suitable for driving a sump pump or opening a drain valve, etc.

NATIONAL SEMICONDUCTOR

Fig. 13-13

LIQUID-LEVEL MONITOR

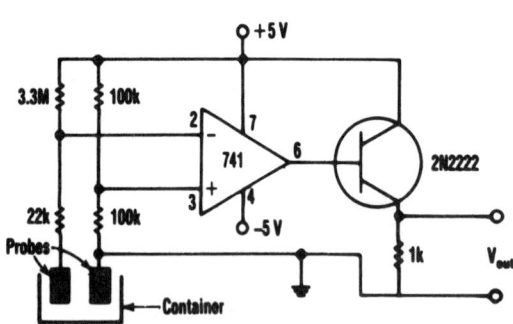

ELECTRONIC DESIGN   *Fig. 13-14*

This monitor uses a common 741 amp configured as a comparator and a low-cost npn transistor as an output driver. With no liquid detected, a voltage of about 2.92 V is present in the op amp's inverting input at pin 2. The 100-kΩ resistors establish a reference voltage of +2.5 V at the noninverting input at pin 3 of the op amp. Under those conditions, the op amp's output is −3.56 V, which keeps the 2N2222 transistor turned off and the voltage across its 1-kΩ output load resistor at 0 V. When liquid reaches the probes, the 3.3-MΩ and 22-kΩ resistor circuit conductively connects to ground. When enough current, about 1.4 μA, flows through the liquid, the small 30-mV drop developed across the 22-kΩ resistor drives the op amp to deliver an output voltage of about 4.42 V. This voltage then drives a 2N2222 transistor into saturation, which generates a voltage drop of about 3.86 V across its 1-kΩ output load resistor.

WATER-LEVEL INDICATOR

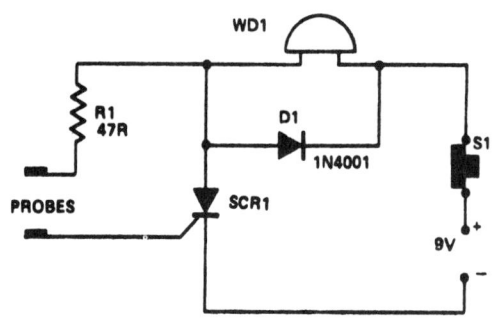

ELECTRONICS TODAY INTERNATIONAL   *Fig. 13-15*

In this warning device, WD1 is in series with SCR1. When the liquid level causes a conductive path between the probes, the SCR conducts sounding WD1. The warning device can be a Sonalert™, a lamp or a buzzer. D1 acts as a transient suppressor. Press S1 to reset the circuit.

WATER-LEVEL ALARM

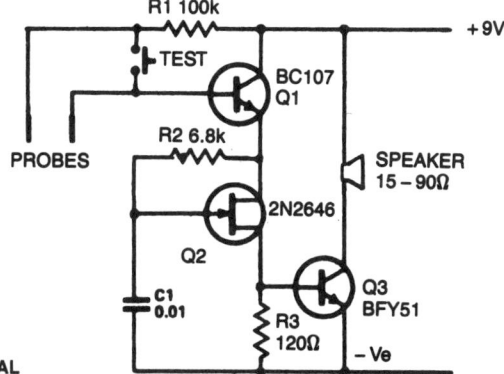

Fig. 13-16

The circuit draws so little current that the shelf-line of the battery is the limiting factor. The only current drawn is the leakage of the transistor. The circuit is shown in the form of a water-level alarm, but by using different forms of probes, it can act as a rain alarm or shorting alarm; anything from zero to about 1 MΩ between the probes will trigger it. Q1 acts as a switch, which applies current to the unijunction relaxation oscillator Q2. The alarm signal frequency is controlled by values and ratios of C_1/R_2. Pulses switch Q3 on and off, applying a signal to the speaker. Almost any npn silicon transistor can be used for Q1 and Q3, and almost any unijunction can be used for Q2.

LOW-LEVEL WARNING WITH AUDIO OUTPUT

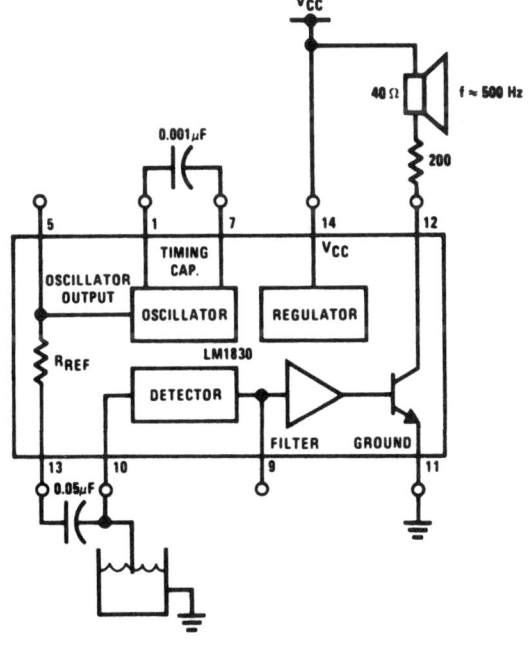

Fig. 13-17

14

Metal Detectors

The sources of the following circuits are contained in the Sources section, which begins on page 214. The figure number in the box of each circuit correlates to the source entry in the Sources section.

Micropower Metal Detector
Simple Treasure Locator
Metal Locator I
Metal Locator II

MICROPOWER METAL DETECTOR

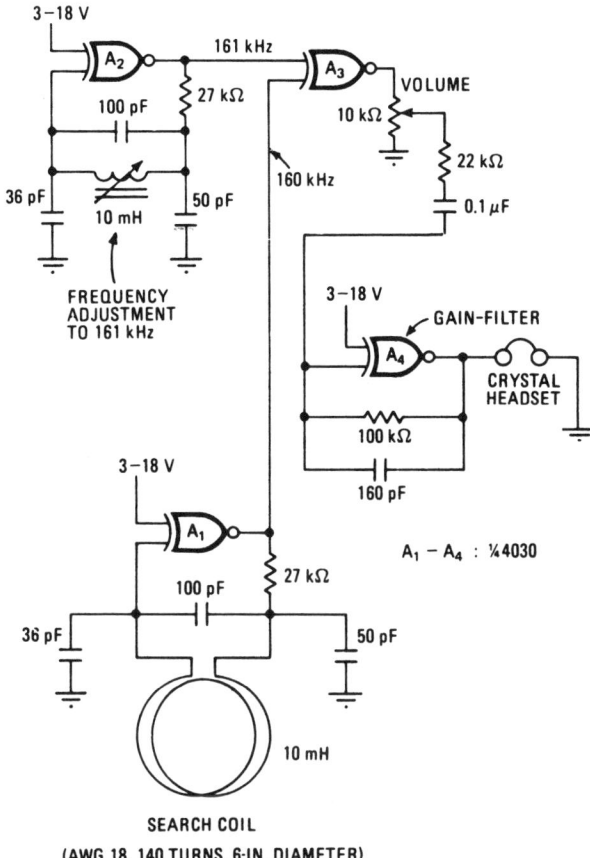

ELECTRONICS

Fig. 14-1

This battery-powered metal detector uses four exclusive-OR gates contained in the 4030 CMOS integrated circuit. The gates are wired as twin-oscillators and a search coil serves as the inductance element in one of the oscillators. When the coil is brought near metal, the resultant change in its effective inductance changes the oscillator's frequency. Gates A1 and A2 form the two oscillators, which are turned to 160 and 161 kHz respectively. The pulses produced by each oscillator are mixed in A3, its output contains sum and difference frequencies at 1 and 321 kHz. The 321-kHz signal is filtered out by the 10-kHz low-pass filter at A4, leaving the 1-kHz signal to be amplified for the crystal headset (connected at the output). The device's sensitivity is sufficient to detect coin-sized objects a foot away.

SIMPLE TREASURE LOCATOR

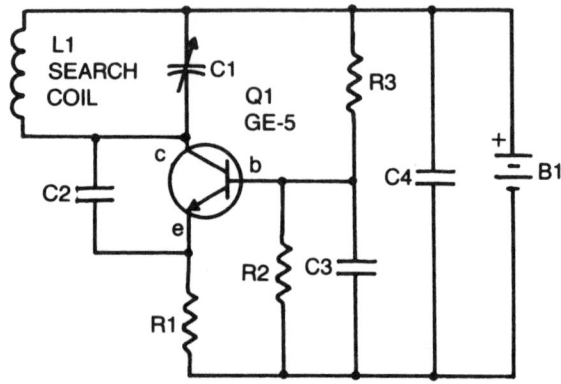

PARTS LIST FOR
LO-PARTS TREASURE LOCATOR
B1—9-Vdc transistor battery
C1—365-pF trimmer or variable capacitor
C2—100-pF, 100-V silver mica capacitor
C3—0.05-μF, disc capacitor
C4—4.7- or 5-μF, 12-V electrolytic capacitor
L1—Search coil consisting of 18 turns of #22 enamel wire
    scramble wound on 4-in. diameter form
Q1—RCA SK3011 npn transistor or equiv.
R1—680-ohm, ½-watt resistor
R2—10,000-ohm, ½-watt resistor
R3—47,000-ohm, ½-watt resistor

101 ELECTRONIC PROJECTS　　　　　　　　　*Fig. 14-2*

This locator uses a transistor radio as the detector. With the radio tuned to a weak station, adjust C1 so that the locator oscillator beats against the received signal. When the search head passes over metal, the inductance of L1 changes and thereby changes the locator oscillator's frequency and changes the beat tone in the radio. The search coil consists of 18 turns of #22 enameled wire scramble wound on a 4″-diameter form. After the coil is wound and checked for proper operation, saturate the coil with RTV adhesive for stable operation of the locator.

METAL LOCATOR I

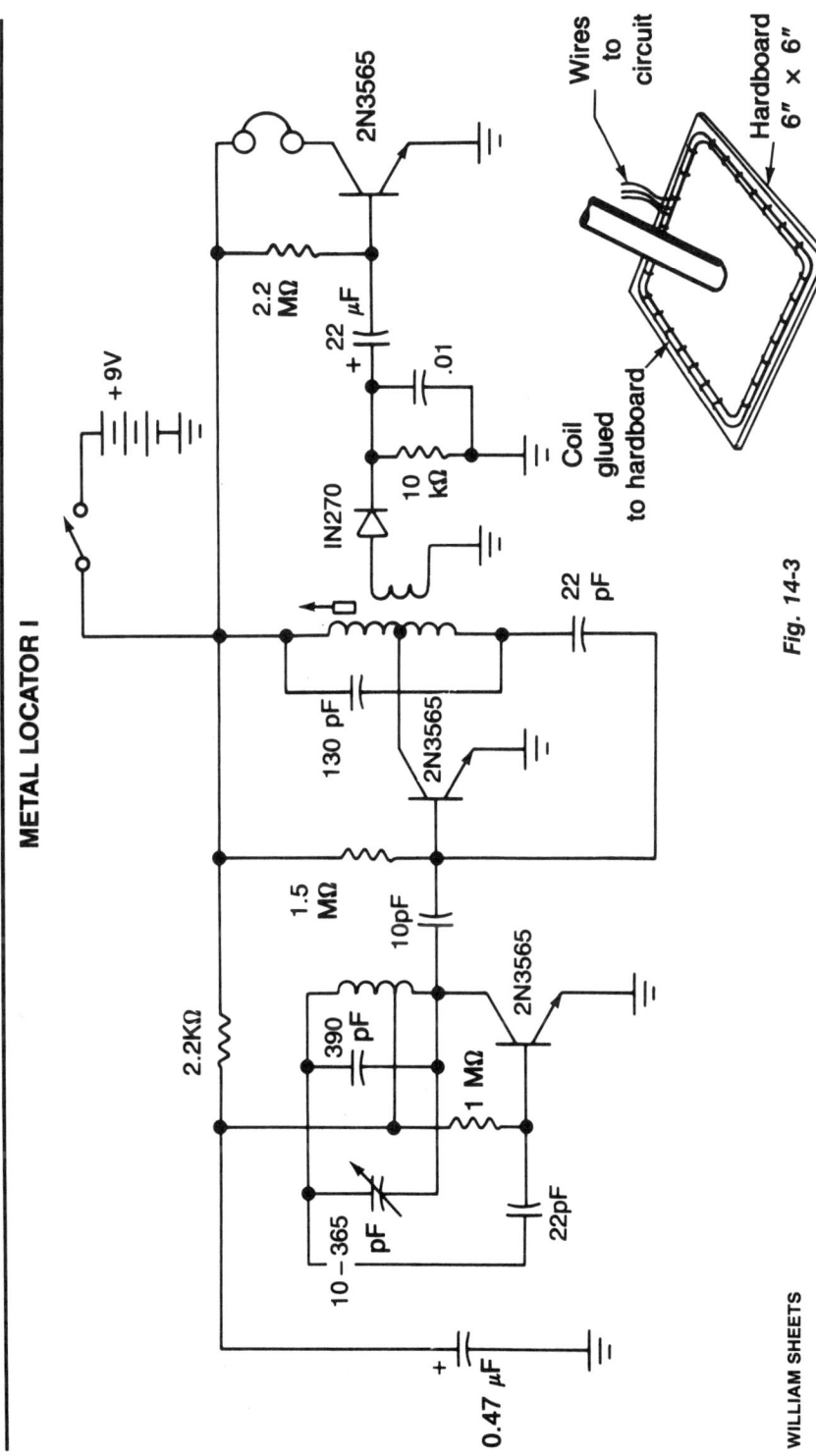

Fig. 14-3

WILLIAM SHEETS

The circuit consists of two oscillators, both of which work at about 465 kHz. One uses an inductor (the search coil L1). The oscillators are coupled by a capacitor (10 pF). A beat note (produced if the two oscillators are working closely together) is detected by the diode and fed to the headphone amplifier and the 22-μF capacitor. The search-coil oscillator is tuned by a 10- to 365-pF variable capacitor. The search coil is composed of 22 turns of wire (any gauge between 24 swg and 36 swg enamel) center tapped. The wire should be wound on a temporary form, then taped and glued to a piece of hardboard. The coil size should be about 6″×6″. Headphones should be high impedance.

METAL LOCATOR II

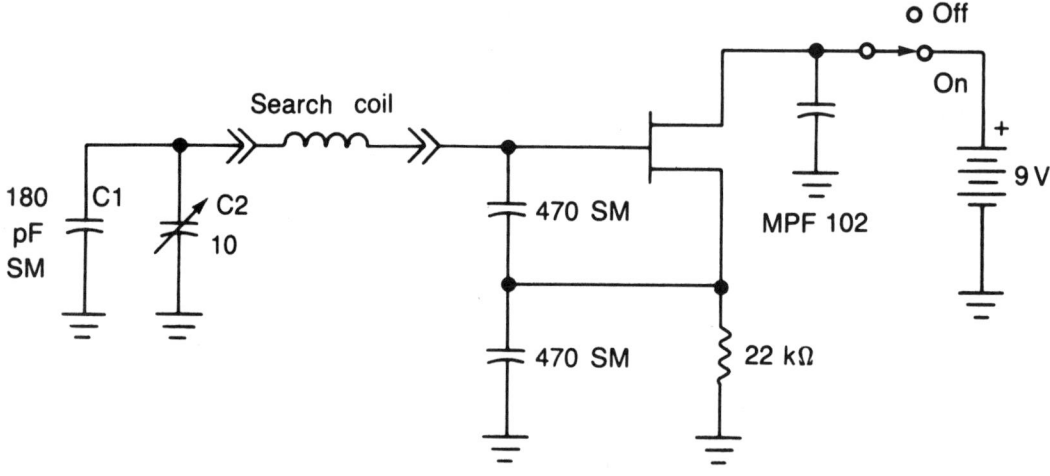

WILLIAM SHEETS

Fig. 14-4

The search coil, C1, and C2 form a turned circuit for the oscillator, which is tuned near the center of the broadcast band. Tune a portable radio to a station near the middle of the band, then tune C2 until a squeal is heard as the two signals mix to produce a beat (heterodyne) note. Metal near the search coil will detune the circuit slightly, changing the pitch of the squeal. The search coil is 20 turns of number 30 enameled wire, wound on a 6″ × 8″ wood or plastic form. It is affixed at the end of a 30″ to 40″ wooden or plastic pole, and connected to the remainder of the metal-detector circuit through a coaxial cable.

15

Miscellaneous Detectors

The sources of the following circuits are contained in the Sources section, which begins on page 214. The figure number in the box of each circuit correlates to the source entry in the Sources section.

Phase Detector with 10-Bit Accuracy
Frequency-Limit Detector
500-Hz Tone Detector
Audio Decibel-Level Detector with Meter Driver
Precision Envelope Detector

Product Detector
Tone Detector
FM Tuner with a Single-Tuned Detector Coil
True rms Detector
Double-Ended Limit Detector

PHASE DETECTOR WITH 10-BIT ACCURACY

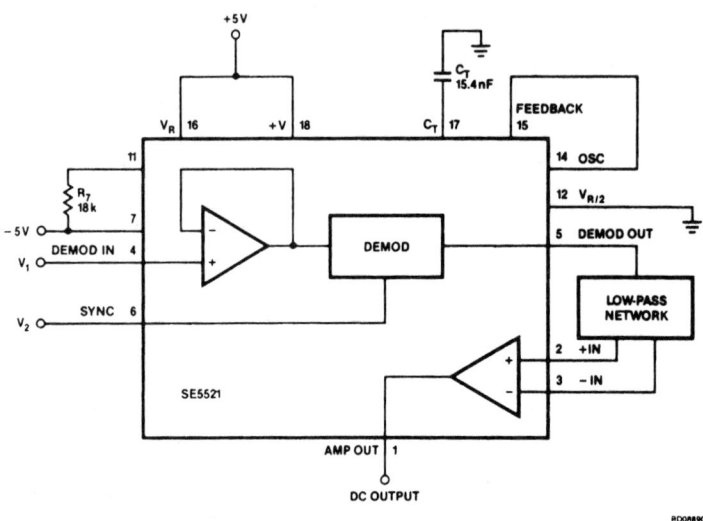

a. Phase Detector Measures Phase Difference Between Signals V₁ and V₂ and Provides dc Output at Pin 1

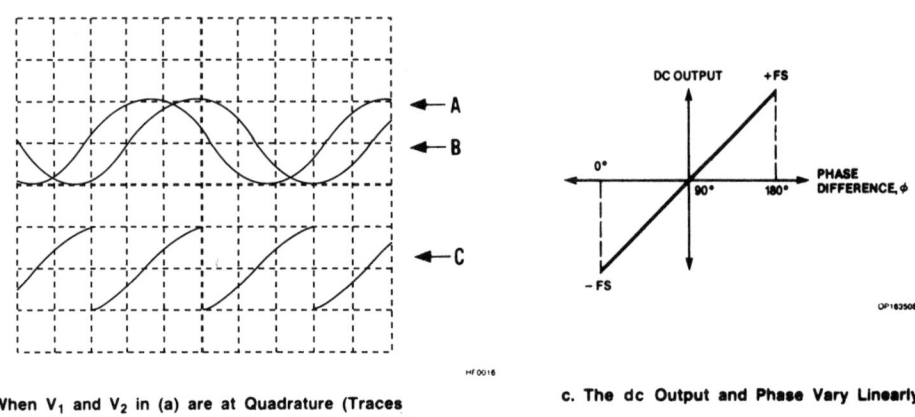

b. When V₁ and V₂ in (a) are at Quadrature (Traces A and B), the DC Component of Demodulator Output (Trace C) is at 0V

c. The dc Output and Phase Vary Linearly

SIGNETICS

Fig. 15-1

Signals of identical frequency are applied to sync input (pin 6) and to the demodulator input (pin 4), respectively, the demodulator functions as a phase detector, and the output dc component is proportional to the phase difference between the two inputs. The signals must be referenced to 0 V for dual-supply operation to $V_R/2$ for single-supply operation. At ±5-V supplies, the demodulator can easily handle 7-V peak-to-peak signals. The low-pass network is configured with the uncommitted amplifier dc output at pin 1 of the device. The dc output is maximum (+full-scale) when V_1 and V_2 are 180° out of phase and minimum (−full-scale) when the signals are in phase.

FREQUENCY-LIMIT DETECTOR

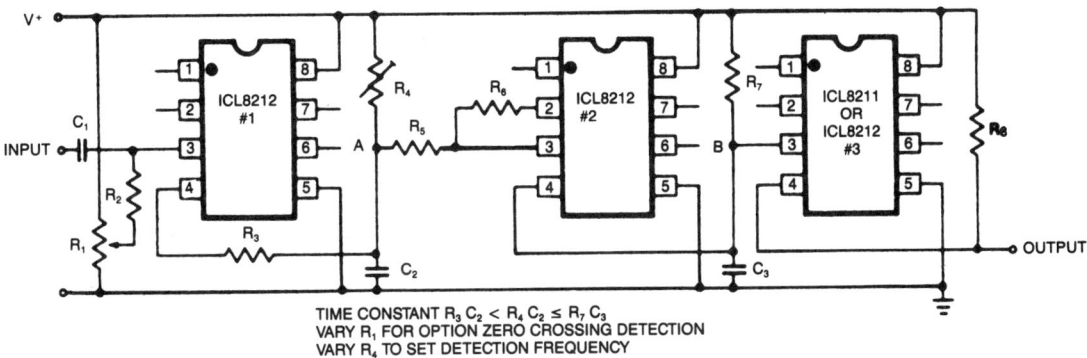

TIME CONSTANT $R_3 C_2 < R_4 C_2 \leq R_7 C_3$
VARY R_1 FOR OPTION ZERO CROSSING DETECTION
VARY R_4 TO SET DETECTION FREQUENCY

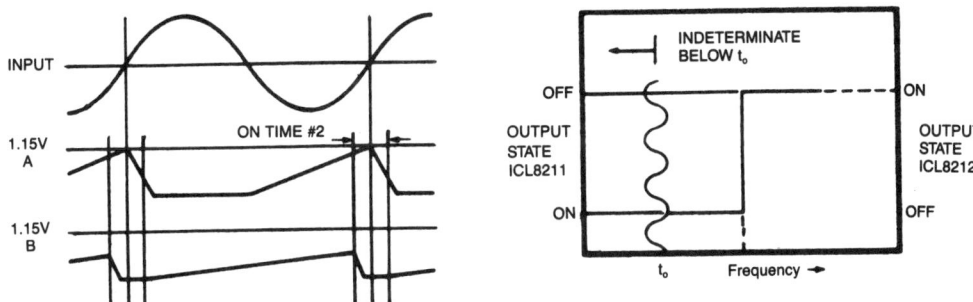

INTERSIL

Fig. 15-2

Simple frequency-limit detectors that provide a go/no-go output for use with varying amplitude input signals can be conveniently implemented with the ICL8211/8212. In the application shown, the first ICL8212 is used as a zero-crossing detector. The output circuit (consisting of R3, R4, and C2) results in a slow-output positive ramp. The negative range is much faster than the positive range. R5 and R6 provide hysteresis so that under all circumstances the second ICL8212 is turned on for sufficient time to discharge C3. The time constant of R7C3 is much greater than R4C2. Depending on the desired output polarities for low- and high-input frequencies, either an ICL8211 or an ICL8212 can be used as the output driver.

The circuit is sensitive to supply voltage variations and should be used with a stabilized power supply. At very low frequencies, the output will switch at the input frequency.

500-Hz TONE DETECTOR

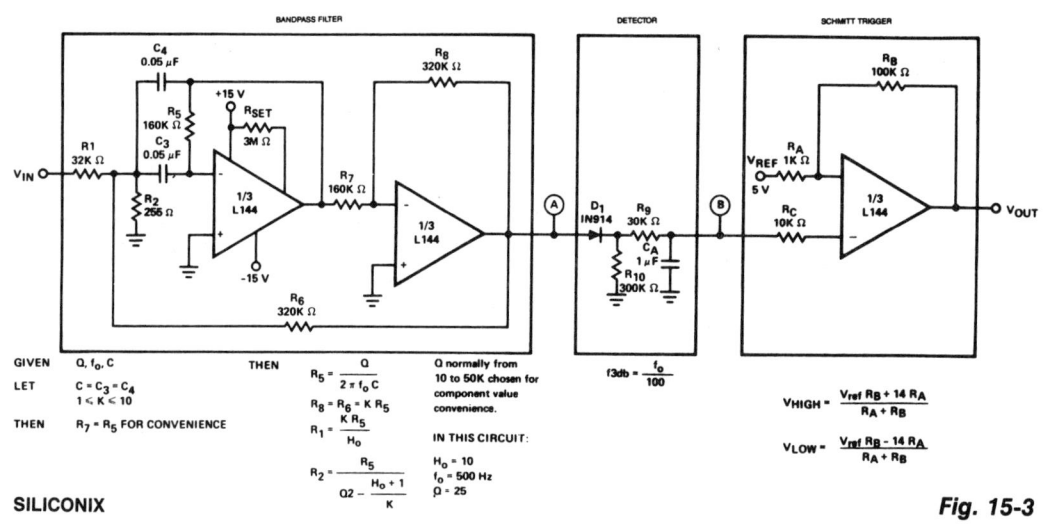

GIVEN   Q, f_o, C   THEN

LET $C = C_3 = C_4$

     $1 < K < 10$

THEN $R_7 = R_5$ FOR CONVENIENCE

$$R_5 = \frac{Q}{2\pi f_o C}$$

$$R_8 = R_6 = K R_5$$

$$R_1 = \frac{K R_5}{H_o}$$

$$R_2 = \frac{R_5}{Q2 - \frac{H_o + 1}{K}}$$

Q normally from 10 to 50K chosen for component value convenience.

IN THIS CIRCUIT:

$H_o = 10$
$f_o = 500$ Hz
$Q = 25$

$$f3db = \frac{f_o}{100}$$

$$V_{HIGH} = \frac{V_{ref} R_B + 14 R_A}{R_A + R_B}$$

$$V_{LOW} = \frac{V_{ref} R_B - 14 R_A}{R_A + R_B}$$

SILICONIX

Fig. 15-3

AUDIO DECIBEL-LEVEL DETECTOR WITH METER DRIVER

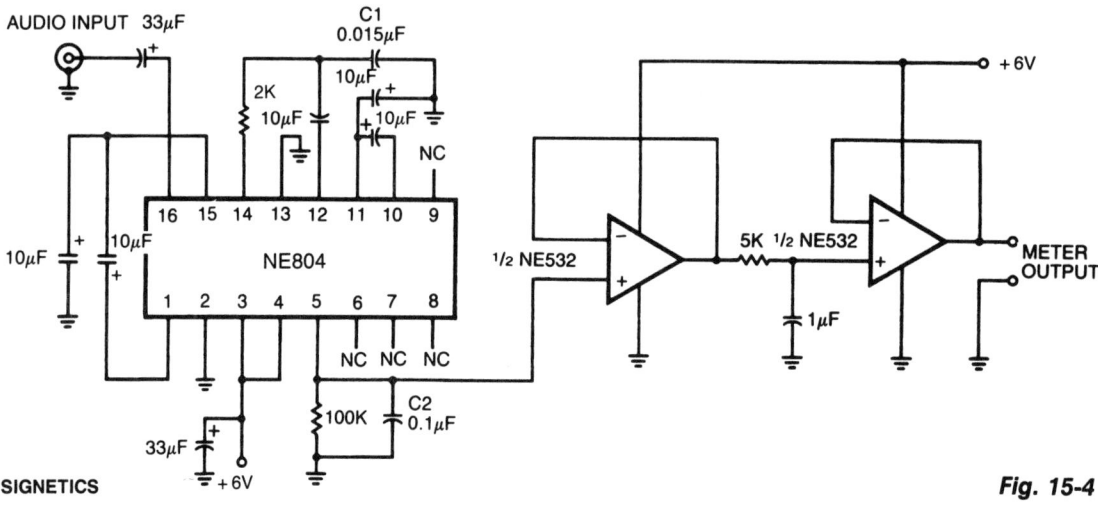

SIGNETICS

Fig. 15-4

This circuit draws very little power, less than 5 mA with a single 6-V power supply, which makes it ideal for portable battery-operated equipment. The small size and low power consumption belie the 90-dB dynamic range and 10.5-μV sensitivity. The dc output voltage is proportional to the $\log_{10}$ of the input signal level. Thus, a standard 0 to 5 voltmeter can be linearly calibrated in decibels over a single 80-dB range. The circuit is within 1.5-dB tolerance over the 80-dB range for audio frequencies from 100 Hz to 10 kHz. Higher audio levels can be measured by placing an attenuator ahead of the input capacitor.

PRECISION ENVELOPE DETECTOR

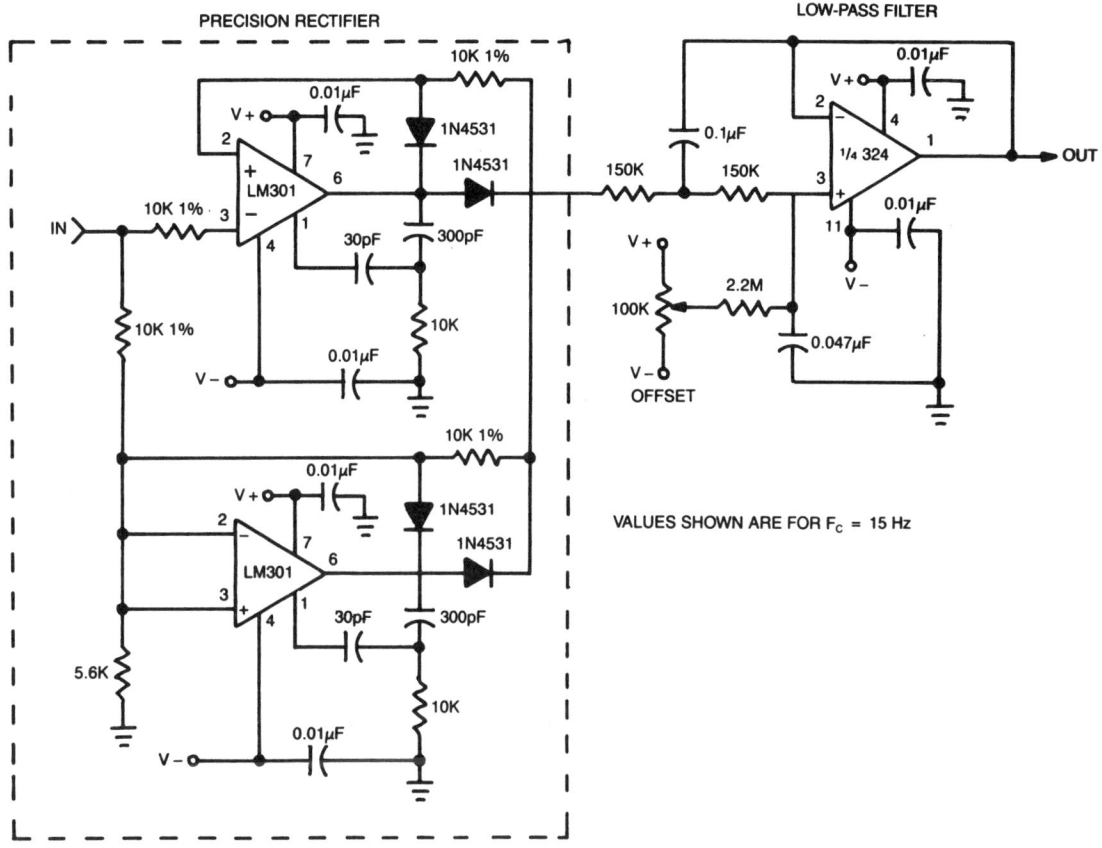

PRECISION RECTIFIER

LOW-PASS FILTER

VALUES SHOWN ARE FOR F$_c$ = 15 Hz

EDN

Fig. 15-5

This circuit is useful for signal-processing sonar data recorded on an instrumentation-quality analog tape recorder. The envelope detector utilizes readily available parts, and furnishes accuracy beyond 100 kHz. Two LM301 op amps connected as precision absolute-value circuits use 2-pole frequency compensation for increased slew rate. And one section of an LM324 quad op amp connected in a Butterworth LPF configuration subjects the rectifier's output to a low-pass filter.

$$f_c = 1/2\pi RC_1$$
$$C_2 = (1/2)C_1$$

PRODUCT DETECTOR

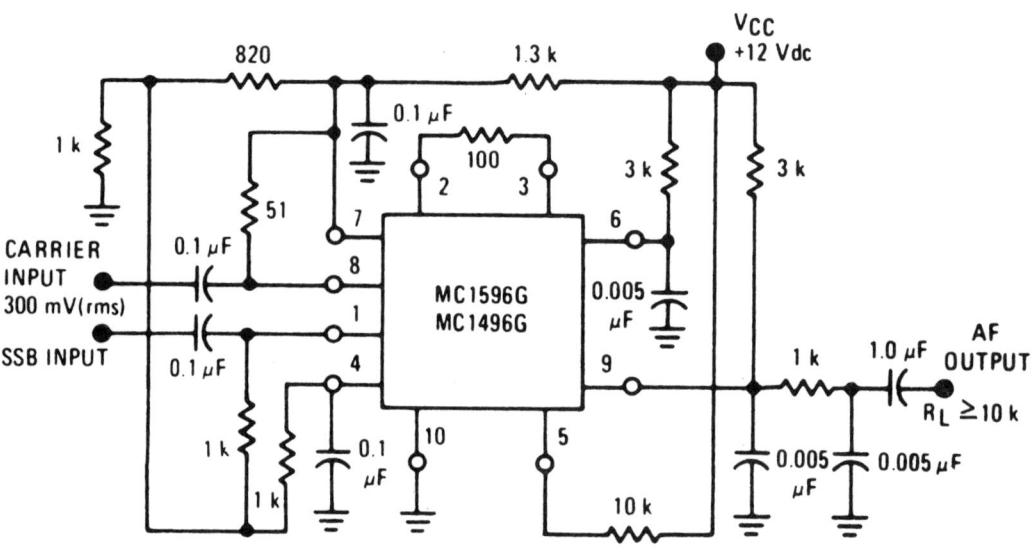

MOTOROLA

Fig. 15-6

The MC1596/MC1496 makes an excellent SSB product detector. This product detector has a sensitivity of 3.0 μV and a dynamic range of 90 dB when operating at an intermediate frequency of 9 MHz. The detector is broadband for the entire high-frequency range. For operation at very low intermediate frequencies down to 50 kHz, the 0.1-μF capacitors on pins 7 and 8 should be increased to 1.0 μF. Also, the output filter at pin 9 can be tailored to a specific intermediate frequency and audio amplifier input impedance. The emitter resistance between pins 2 and 3 can be increased or decreased to adjust circuit gain, sensitivity, and dynamic range. This circuit can also be used as an AM detector by introducing carrier signal at the carrier input and an AM signal at the SSB input. The carrier signal can be derived from the intermediate frequency signal or it can be generated locally. The carrier signal can be introduced with or without modulation, provided that its level is sufficiently high to saturate the upper quad differential amplifier. If the carrier signal is modulated, a 300-mV (rms) input level is recommended.

TONE DETECTOR

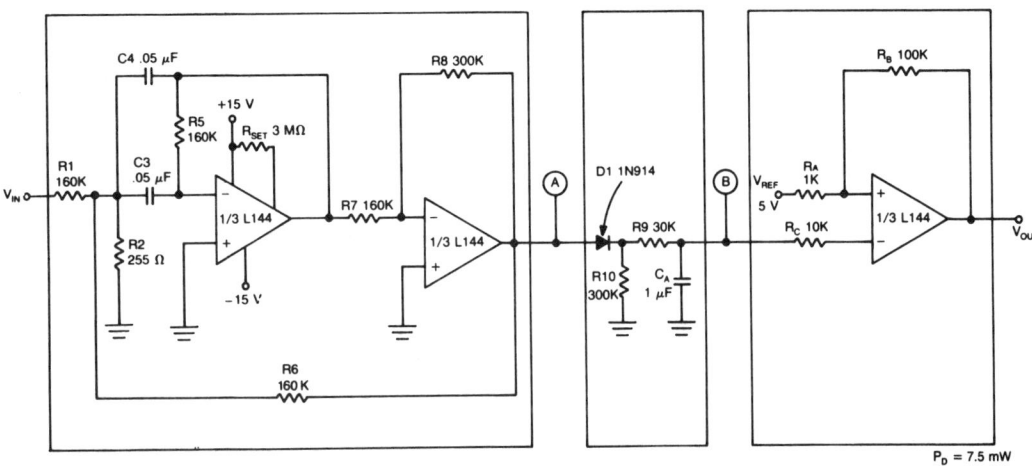

SILICONIX

Fig. 15-7

The detector circuit is made up of a two-amplifier multiple feedback bandpass filter followed by an ac-to-dc detector section and a Schmitt trigger. The bandpass filter (with a Q of greater than 100) passes only 500-Hz inputs, which are in turn rectified by D1 and filtered by R9 and C_A. This filtering action in combination with the trigger level of 5 V for the Schmitt device ensures that at least 55 cycles of 500-Hz input must be present before the output will react to a tone input.

FM TUNER WITH A SINGLE-TUNED DETECTOR COIL

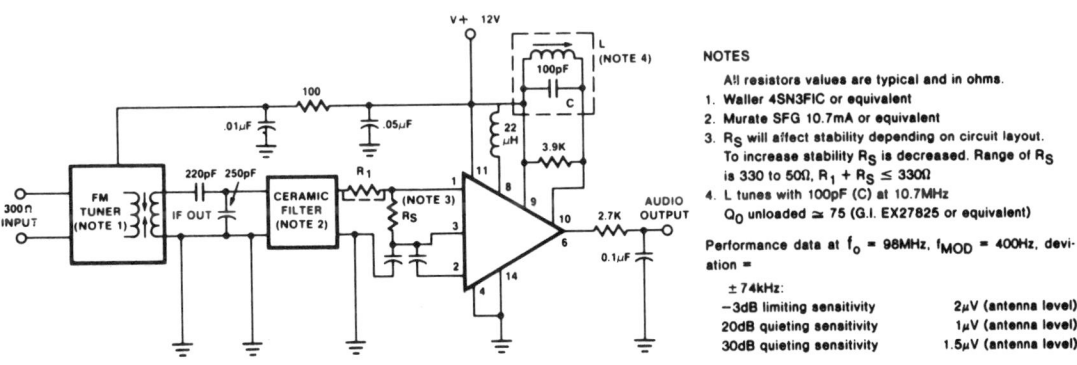

NOTES

All resistors values are typical and in ohms.
1. Waller 4SN3FIC or equivalent
2. Murate SFG 10.7mA or equivalent
3. R_S will affect stability depending on circuit layout. To increase stability R_S is decreased. Range of R_S is 330 to 50Ω, $R_1 + R_S \leq 330Ω$
4. L tunes with 100pF (C) at 10.7MHz
   Q_O unloaded $\simeq$ 75 (G.I. EX27825 or equivalent)

Performance data at f_o = 98MHz, f_{MOD} = 400Hz, deviation =

±74kHz:

−3dB limiting sensitivity	2μV (antenna level)
20dB quieting sensitivity	1μV (antenna level)
30dB quieting sensitivity	1.5μV (antenna level)

SIGNETICS

Fig. 15-8

TRUE rms DETECTOR

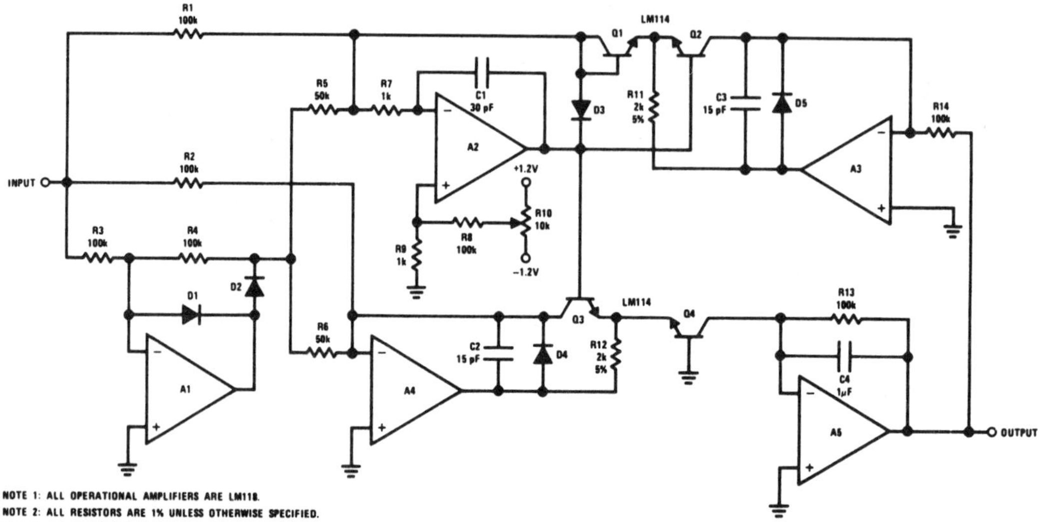

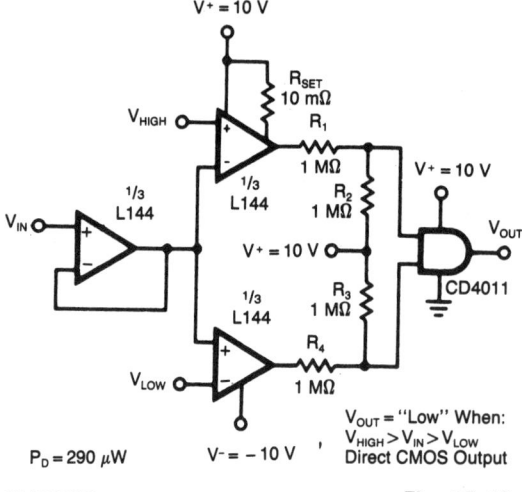

NOTE 1: ALL OPERATIONAL AMPLIFIERS ARE LM118.
NOTE 2: ALL RESISTORS ARE 1% UNLESS OTHERWISE SPECIFIED.
NOTE 3: ALL DIODES ARE 1N914.
NOTE 4: SUPPLY VOLTAGE ±15V.

NATIONAL SEMICONDUCTOR

Fig. 15-9

The circuit will provide a dc output equal to the rms value of the input. Accuracy is typically 2% for a 20 V_{PP} input signal from 50 Hz to 100 kHz, although it's usable to about 500 kHz. The lower frequency is limited by the size of the filter capacitor. Because the input is dc coupled, it can provide the true rms equivalent of a dc and ac signal.

DOUBLE-ENDED LIMIT DETECTOR

This detector uses three sections of an L144 and a CMOS NAND gate to make a very low power voltage monitor. The 1-MΩ resistors (R1, R2, R3, and R4) translate the bipolar ±10-V swing of the op amps to a 0- to 10-V swing that is acceptable to the ground-referenced CMOS logic. The total power dissipation is 290 μW while in limit and 330 μW while out of limit.

V_{OUT} = "Low" When:
$V_{HIGH} > V_{IN} > V_{LOW}$
Direct CMOS Output

P_D = 290 μW

SILICONIX

Fig. 15-10

16

Moisture Detectors

The sources of the following circuits are contained in the Sources section, which begins on page 214. The figure number in the box of each circuit correlates to the source entry in the Sources section.

Soil-Moisture Meter   Automatic Plant Waterer
Moisture Detector   Plant-Water Monitor
Plant-Water Gauge   Low-Cost Humidity Sensor

SOIL-MOISTURE METER

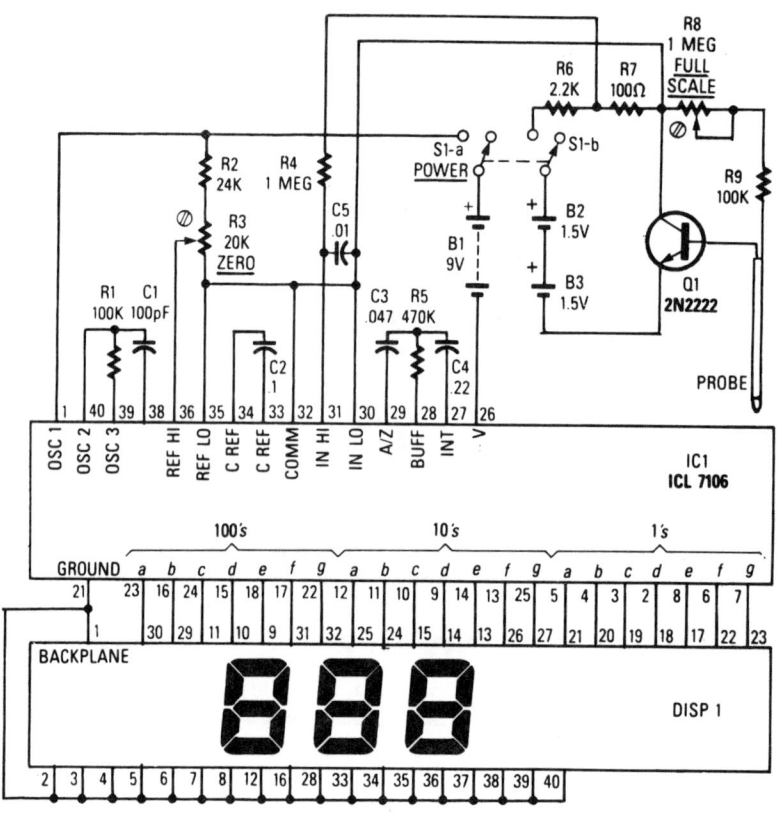

Fig. 16-1

IC1, an Intersil ICL7106, contains an a/d converter, a 3 1/2-digit LCD driver, a clock, a voltage reference, seven segment decoders, and display drivers. A similar part, the ICL7107, can be used to drive seven segment LEDs. The probe body is a five-inch length of light-weight aluminum tubing. The leads from the circuit are connected to the body and tip of the probe. The sensor functions as a variable resistor that varies Q1's base current, hence its collector current. The varying collector current produces a varying voltage across 100-Ω resistor R7, and that voltage is what IC1 converts for display.

The LCD consumes about 25 μA, and IC1 consumes under 2 mA, so the circuit will run for a long time when it is powered by a standard 9-V battery. The current drain of the two 1.5-V AA cells is also very low: under 300 μA.

To calibrate, rotate R3 to the center of its range. Then place the end of the probe into a glass of water and adjust R8 for a reading of 100. When you remove the probe from the water, the LCD should indicate 000. You might have to adjust R3 slightly for the display to indicate 000. If so, readjust R8 with the probe immersed. Check for a reading of 000 again with the probe out of the water.

MOISTURE DETECTOR

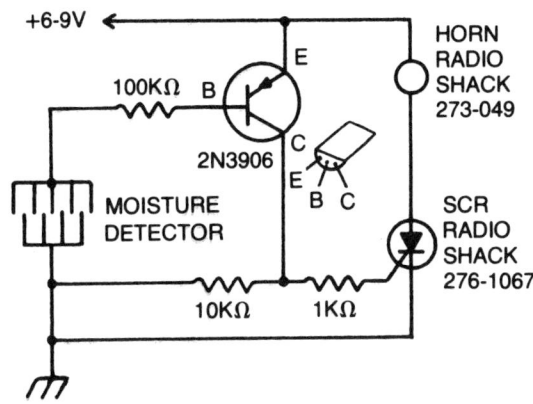

The detector is made of fine wires spaced about one or two inches apart. When the area between a pair of wires becomes moist, the horn will sound. To turn if off, dc power must be disconnected.

MODERN ELECTRONICS

Fig. 16-2

PLANT-WATER GAUGE

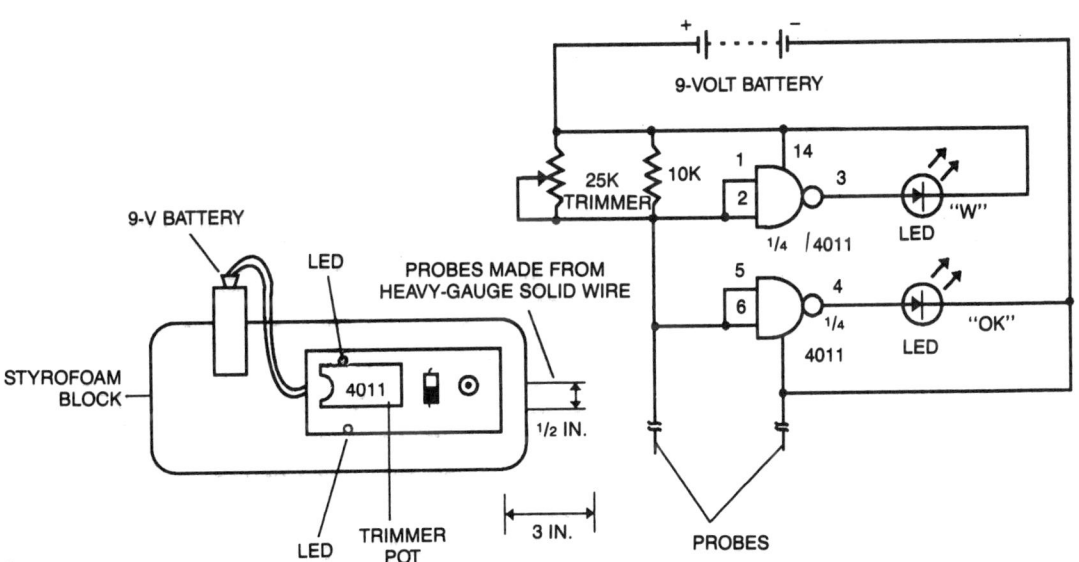

RADIO-ELECTRONICS

Fig. 16-3

To calibrate the gauge, connect the battery and press the probes gently into a pot containing a plant that is just on the verge of needing water (stick it in so that only an inch of the probe is left visible at the top). Turn the potentiometer until the ''OK'' LED lights and then turn it back to the point where that LED goes out and the ''Water,'' LED just comes on. The device should now be properly adjusted.

AUTOMATIC PLANT WATERER

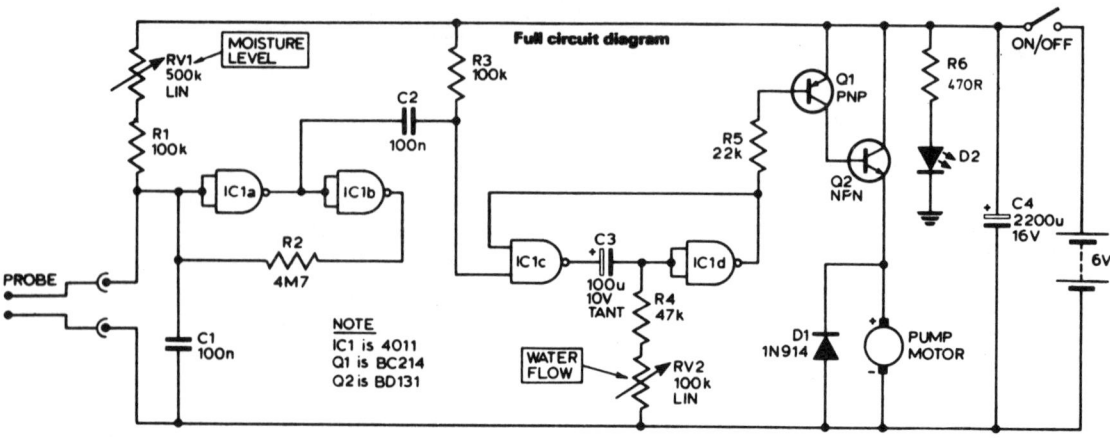

ELECTRONICS TODAY INTERNATIONAL

Fig. 16-4

The unit consists of a sensor, timer, and an electric water pump. The sensor is embedded in the soil, and when dry, the electronics operate the water pump for a preset time. The circuit is composed of a level-sensitive Schmitt trigger, a variable-time monostable, and an output driver. When the resistance across the probe increases beyond a set value (i.e., the soil dries), the Schmitt is triggered. C2 feeds a negative-going pulse to the monostable when the Schmitt triggers and R2 acts as feedback to ensure a fast switching action.

PLANT-WATER MONITOR

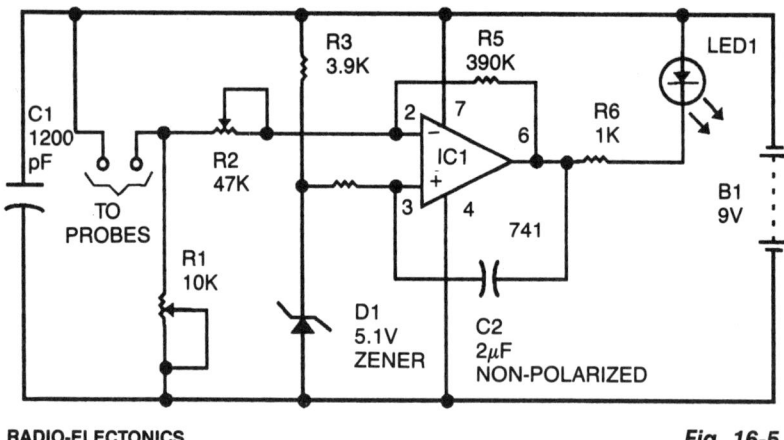

RADIO-ELECTRONICS

Fig. 16-5

When the soil is moist, the LED glows. If the moisture falls below a certain predetermined level, the LED begins to flash. If there is still less moisture, the LED turns off. To calibrate, connect the battery and insert the probe into a container of dry soil. Set R1 to its maximum value, then reduce that resistance until the LED begins to flash. The range over which the LED flashes before going out can be adjusted using R2.

LOW-COST HUMIDITY SENSOR

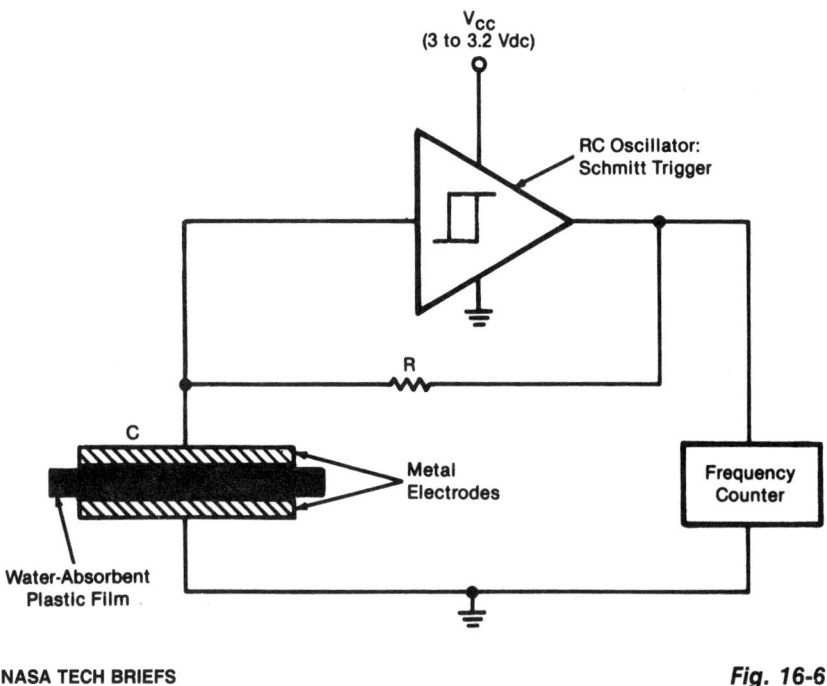

NASA TECH BRIEFS

Fig. 16-6

The sensor is an RC oscillator in which a water-absorbent plastic film is the insulator in the capacitive element. The capacitance of the film increases with the amount of water it absorbs from the air, and thus reduces the oscillation frequency. A frequency counter produces a digital output that represents the change in frequency and hence the change in relative humidity. The sensor can be used to measure humidity in the atmosphere, in the soil, and in industrial gases, for example. A Schmitt-trigger-type IC is connected to the capacitor, which consists of a film of a commercially produced sulfonated fluorocarbon polymer, 2″ (5.08 cm) square, sandwiched between perforated metal plates. The oscillation frequency decreases almost linearly from about 100 Hz to 16 kHz as the relative humidity increases from about 20 to 76%.

17

Null Detector

The source of the following circuit is contained in the Sources section, which begins on page 214. The figure number contained in the box of the circuit correlates to the source entry in the Sources section.

Null Detector

NULL DETECTOR

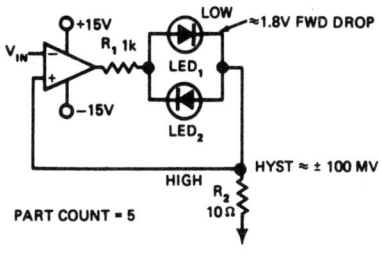

PART COUNT = 5

In this indicating comparator circuit, R2 sets the hysteresis. If the 741 saturates at ± 12 V, the current in R1 will be approximately ± 10 mA if 0.1 V hysteresis is desired. Then 0.1 V/10 mA $= 10\Omega = R_2$.

EDN

Fig. 17-1

18

Optical Detectors

The sources of the following circuits are contained in the Sources section, which begins on page 214. The figure number in the box of each circuit correlates to the source entry in the Sources section.

FOUR-QUADRANT PHOTOCONDUCTIVE DETECTOR AMPLIFIER

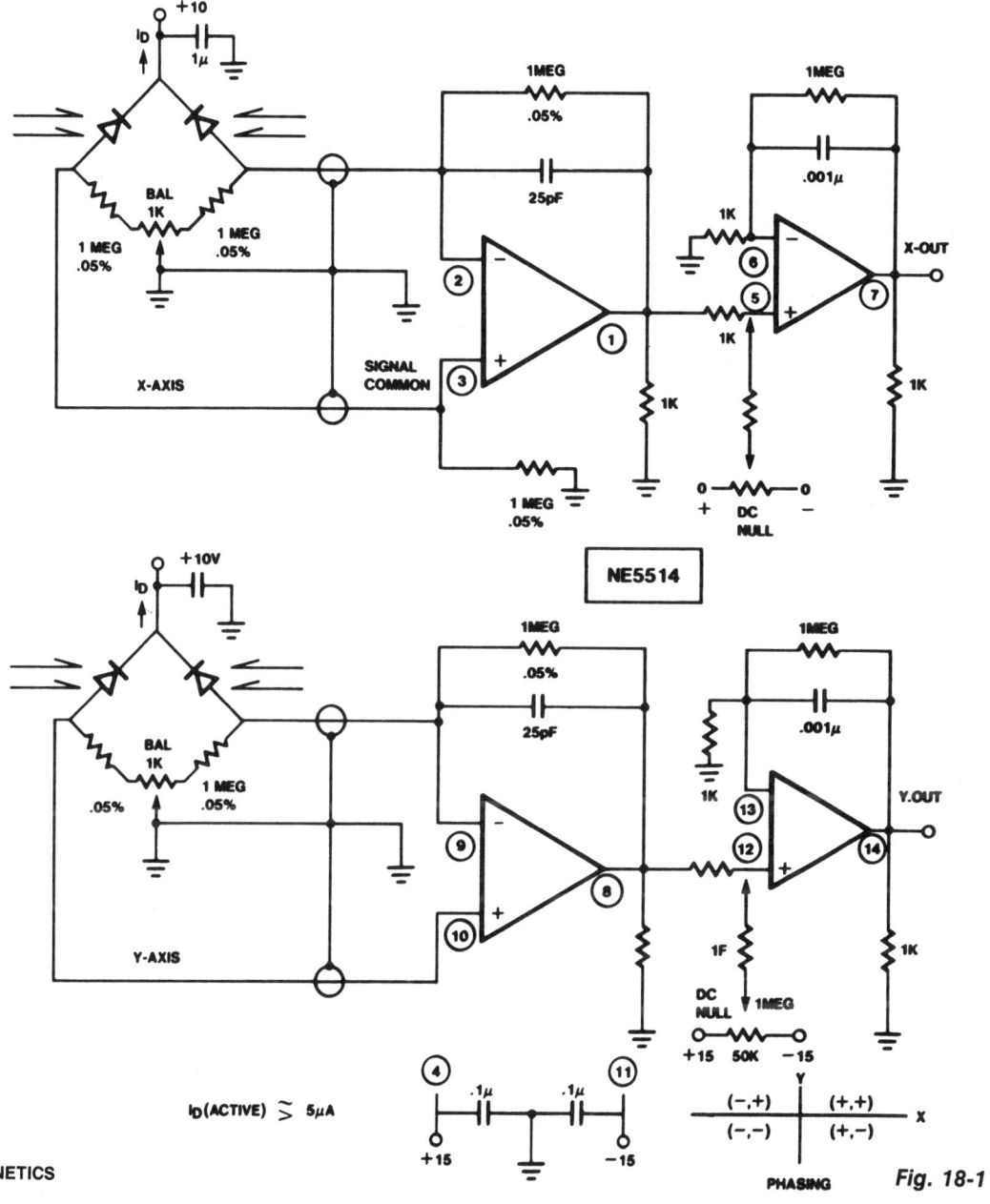

SIGNETICS

Fig. 18-1

Use this circuit to sense four-quadrant motion of a light source. By properly summing the signals from the X and Y axes, four-quadrant output can be fed to an X-Y plotter, oscilloscope, or computer for simulation. IC = NE/SE5514.

OPTICAL SCHMITT TRIGGER

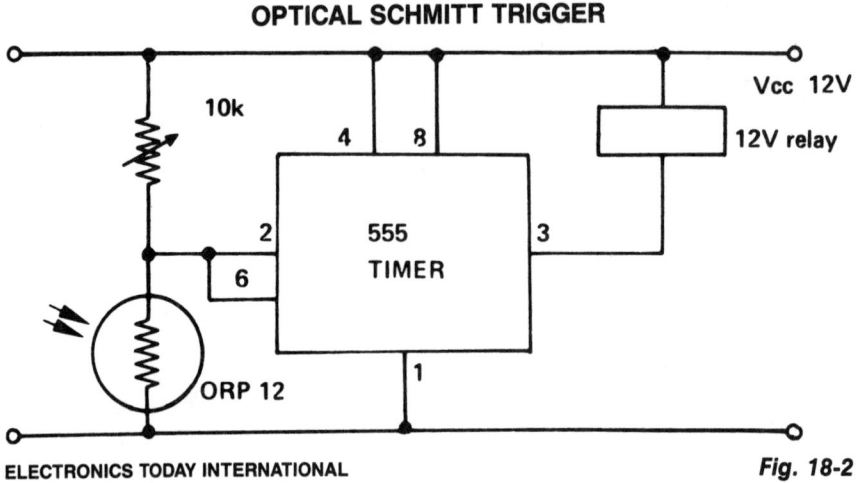

ELECTRONICS TODAY INTERNATIONAL   *Fig. 18-2*

This circuit shows a 555 with its trigger and threshold inputs connected together used to energize a relay when the light level on a photoconductive cell falls below a preset value. This circuit can be used in other applications where a high input impedance and low output impedance are required for minimum component count.

ADJUSTABLE LIGHT-DETECTION SWITCH

COMPUTERS & ELECTRONICS   *Fig. 18-3*

R2 sets the circuit's threshold. When the light intensity at PCI's surface is decreased, the resistance of PC1 (a cadmium-sulfide photoresistor) is increased. This decreases the voltage at the inverting input of the 741. When the reference voltage at the 741's noninverting input is properly adjusted via R2, the comparator will switch from low to high when PC1 is darkened. This turns on Q1, which, in turn, pulls in relay K1.

LIGHT-INTERRUPTION DETECTOR

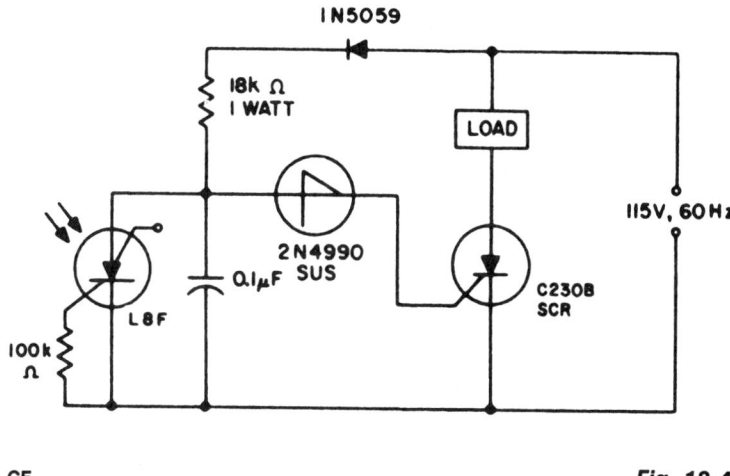

GE

Fig. 18-4

When the light incident on the LASCR is interrupted, the voltage at the anode to the 2N4990 unilateral switch goes positive on the next positive cycle of the power, which triggers the switch and the C230 SCR when the switching voltage of the unilateral switch is reached. This will cause the load to be energized for as long as light is not incident on the LASCR.

OPTICAL RECEIVER

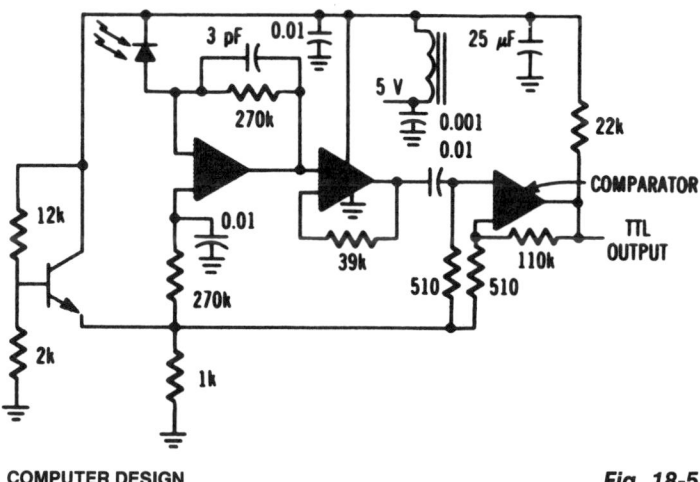

COMPUTER DESIGN

Fig. 18-5

The MFOD1100 PIN diode requires shielding from EMI.

LIGHT-ISOLATED SOLID-STATE POWER-RELAY CIRCUITS

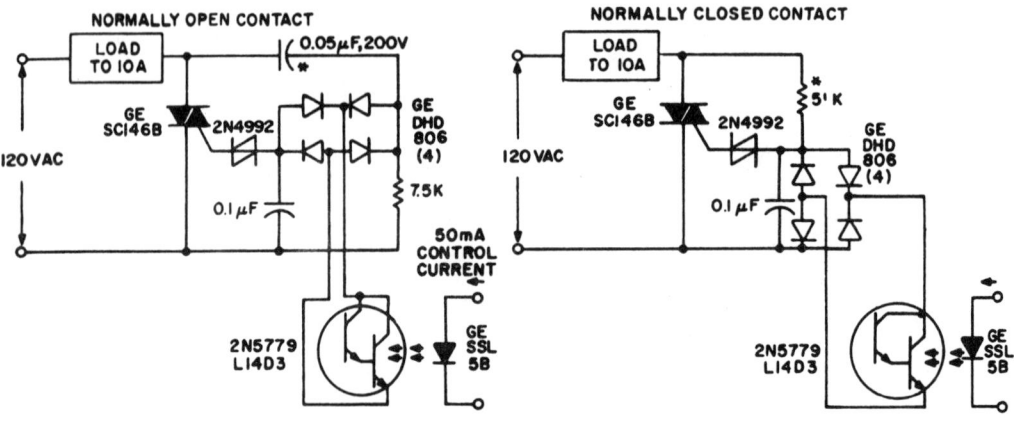

GE

Fig. 18-6

Both circuits use the GE SC146B 200-V 10-A triac as load-current contacts. These triacs are triggered by normal SBS (2N4992) trigger circuits, which are controlled by the photo-Darlington, acting through the DA806 bridge as an ac photo switch. To operate the relays at other line voltages, the asterisked (*) components are scaled to supply identical current. Ratings must be changed as required. Incandescent lamps can be used in place of the LEDs, if desired.

PRECISION PHOTODIODE-LEVEL DETECTOR

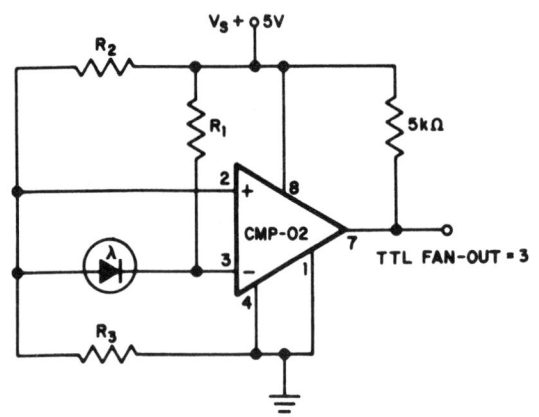

PRECISION MONOLITHICS

Fig. 18-7

For $R_1 = 2.5$ MΩ, $R_2 = R_3 = 5$ MΩ. The output state changes at a photodiode current of 0.5 µA.

LIGHT-OPERATED ON/OFF RELAY

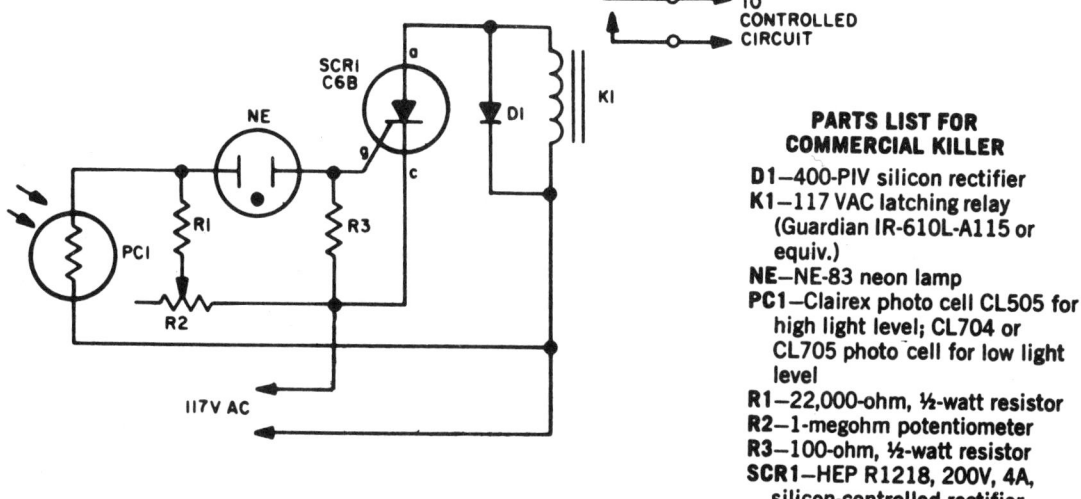

TO
CONTROLLED
CIRCUIT

PARTS LIST FOR COMMERCIAL KILLER

D1—400-PIV silicon rectifier
K1—117 VAC latching relay
(Guardian IR-610L-A115 or
equiv.)
NE—NE-83 neon lamp
PC1—Clairex photo cell CL505 for
high light level; CL704 or
CL705 photo cell for low light
level
R1—22,000-ohm, ½-watt resistor
R2—1-megohm potentiometer
R3—100-ohm, ½-watt resistor
SCR1—HEP R1218, 200V, 4A,
silicon-controlled rectifier

101 ELECTRONIC PROJECTS

Fig. 18-8

When a beam of light strikes the photocell, the voltage across neon lamp NE-1 rises sharply. NE-1 turns on and fires the SCR. K1 is an impulse relay whose contacts stay in position—even after coil current is removed. The first impulse opens K1's contacts, the second impulse closes them, etc.

LOGARITHMIC LIGHT SENSOR

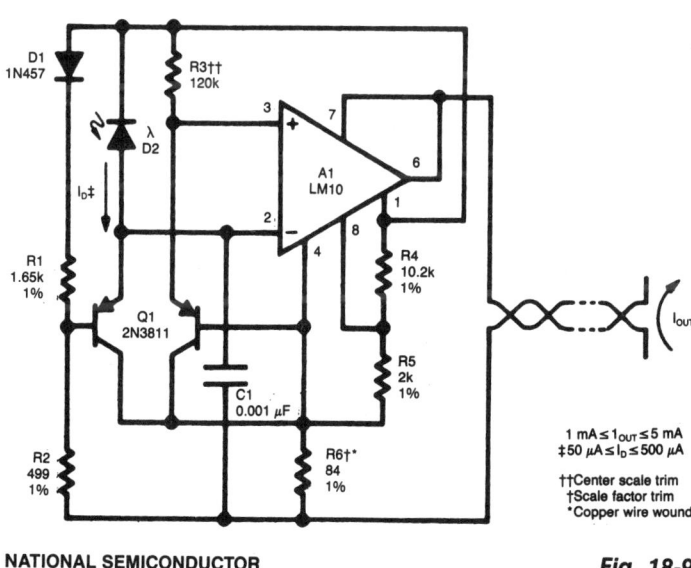

$1\ mA \leq 1_{OUT} \leq 5\ mA$
$\ddagger 50\ \mu A \leq I_D \leq 500\ \mu A$

††Center scale trim
†Scale factor trim
*Copper wire wound

NATIONAL SEMICONDUCTOR

Fig. 18-9

LOW-NOISE INFRARED DETECTOR

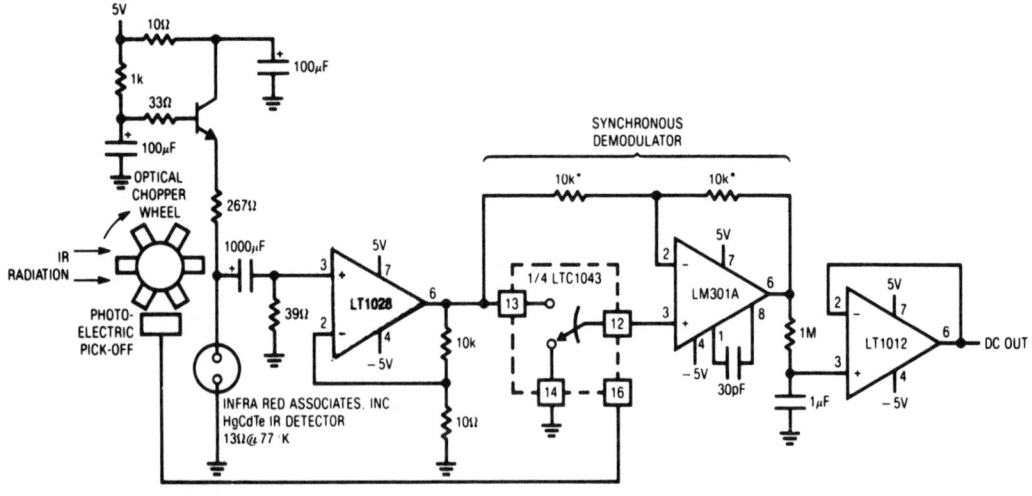

LINEAR TECHNOLOGY

Fig. 18-10

LIGHT-LEVEL SENSOR

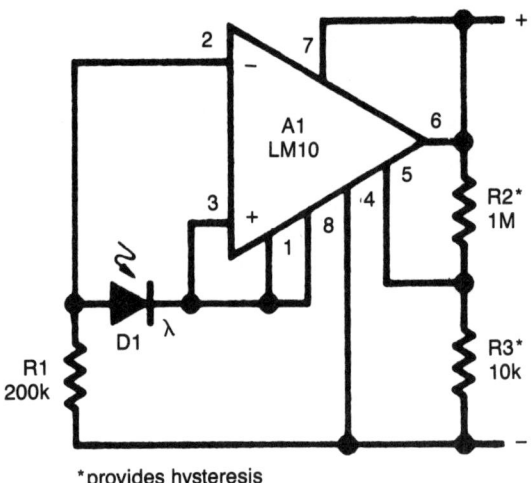

*provides hysteresis

NATIONAL SEMICONDUCTOR

Fig. 18-11

AMBIENT-LIGHT-IGNORING OPTICAL SENSOR

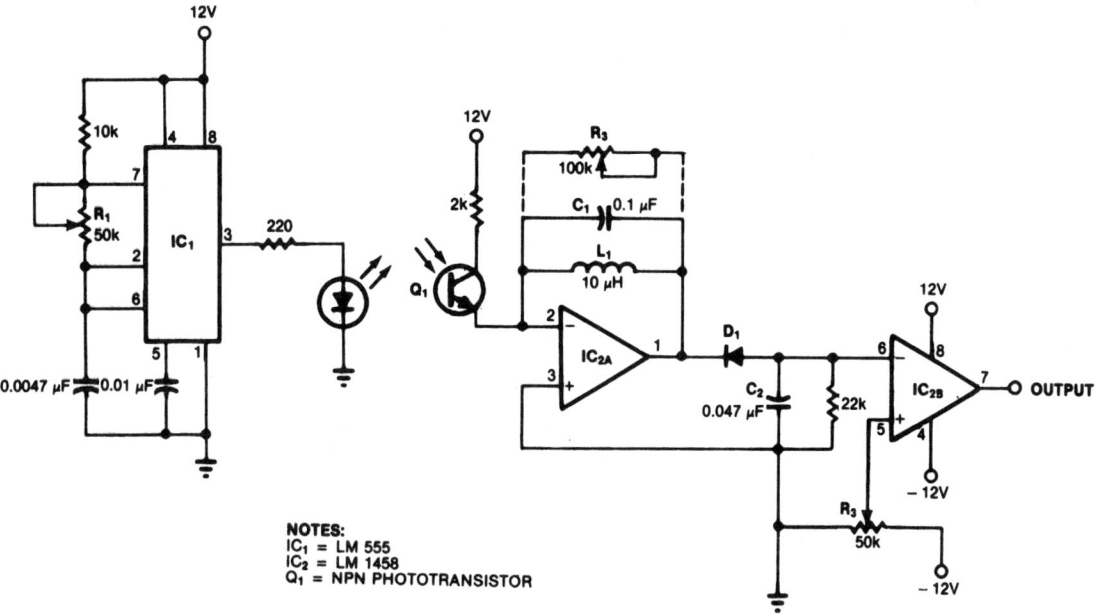

EDN

Fig. 18-12

A resonance-tuned narrow-band amplifier reduces this optical object detector's sensitivity to stray light. C1 and L1 in IC2A's feedback loop cause the op amp to pass only those frequencies at or near the LED's 5-kHz modulation rate. IC2B's output increases when the received signal is sufficient to drop the negative voltage across C2 below the reference set by R2.

19

Overspeed Detectors

The sources of the following circuits are contained in the Sources section, which begins on page 214. The figure number in the box of each circuit correlates to the source entry in the Sources section.

Overspeed Indicator
High-Speed Warning Device
Speed-Warning Device
Speed Alarm

OVERSPEED INDICATOR

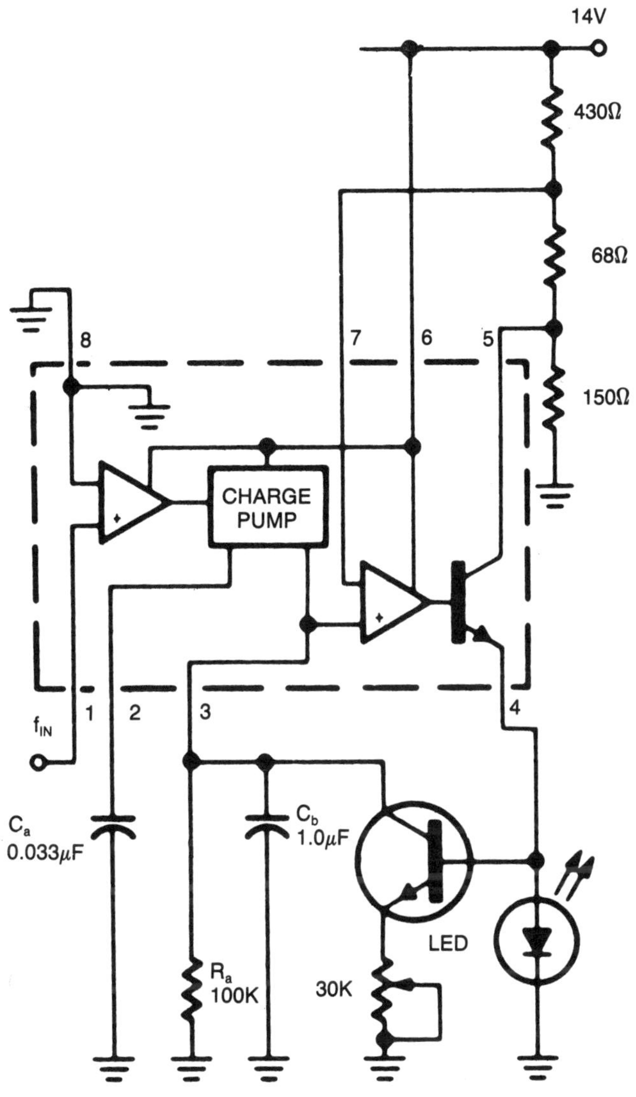

FLASHING BEGINS WHEN f_{IN} – 100 Hz
FLASHING RATE INCREASES WITH INPUT FREQUENCY
INCREASE BEYOND TRIP POINT

NATIONAL SEMICONDUCTOR

Fig. 19-1

An op-amp comparator is used to compare the converter output with a dc threshold voltage. The circuit flashes the LED when the input frequency exceeds 100 Hz. Increases in frequency raise the average current out of terminal 3 so that frequencies above 100 Hz reduce the charge time of C2, increasing the LED flashing rate. IC = LM2907 or LM2917.

HIGH-SPEED WARNING DEVICE

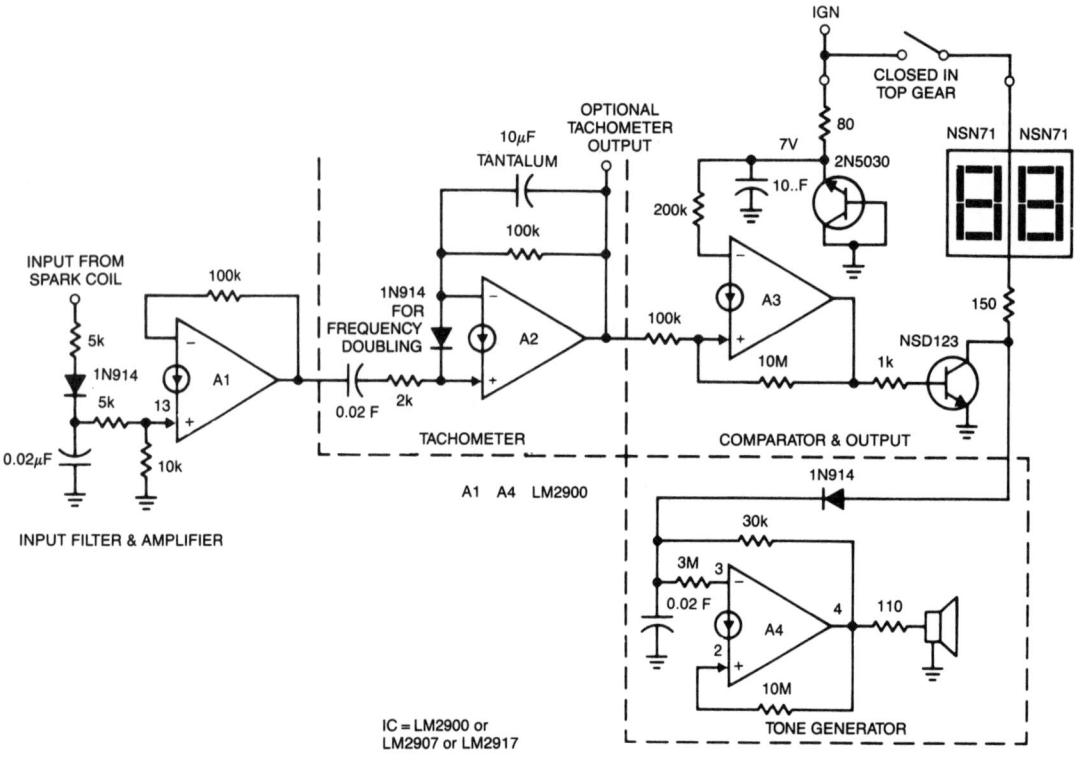

NATIONAL SEMICONDUCTOR

Fig. 19-2

A1 amplifies and regulates the signal from the spark coil. A2 converts frequency to voltage so that its output voltage is proportional to engine rpm. A3 compares the tachometer voltage with the reference voltage and turns on the output transistor at the set speed. Amplifier A4 is used to generate an audible tone whenever the set speed is exceeded.

SPEED-WARNING DEVICE

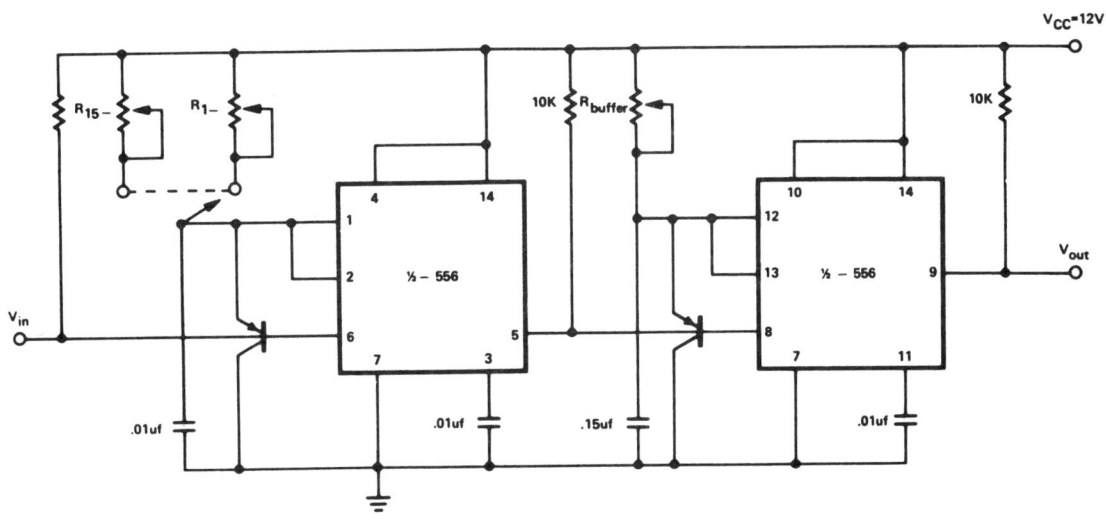

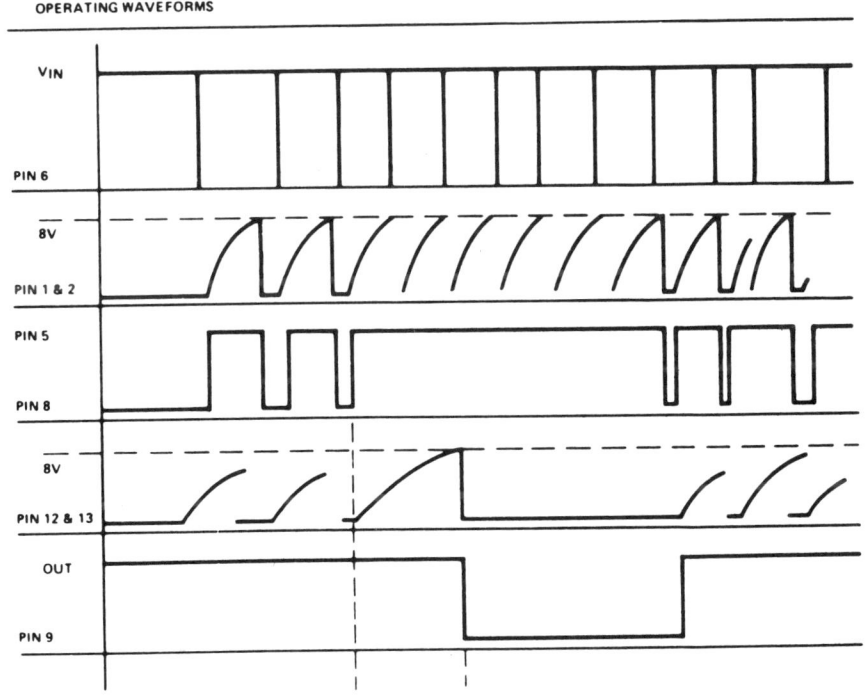

OPERATING WAVEFORMS

Fig. 19-3

SPEED ALARM

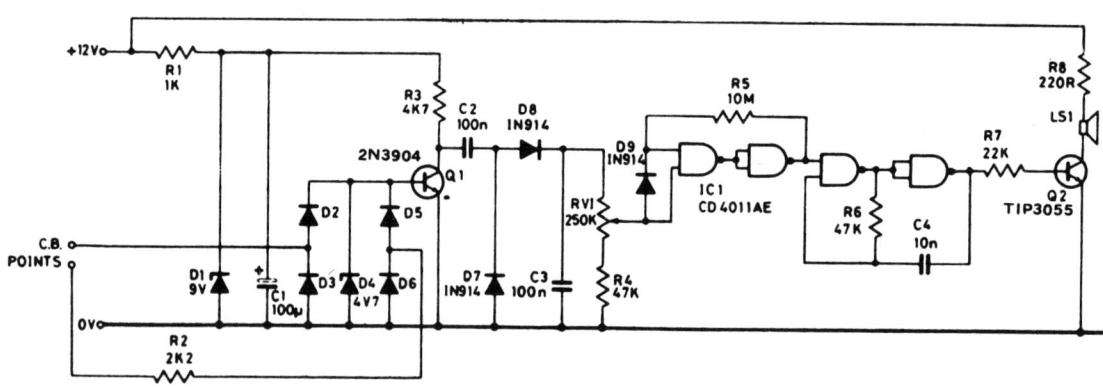

ELECTRONICS TODAY INTERNATIONAL

Fig. 19-4

Pulses from the distributor points are passed through a current-limiting resistor, rectified, and clipped at 4.7 V. Via Q1 and the diode pump, a dc voltage proportional to engine rpm is presented to RV1; the sharp transfer characteristic of a CMOS gate, assisted by feedback, is used to enable the oscillator formed by the remaining half of the 4011. At the preset speed, a nonignorable tone emits from the speaker, and disappears as soon as the speed drops by three or four mph.

20

Overvoltage/Current Detectors

The sources of the following circuits are contained in the Sources section, which begins on page 214. The figure number in the box of each circuit correlates to the source entry in the Sources section.

Overvoltage Protector
High-Speed Electronic Circuit Breaker
12-ns Circuit Breaker
Low-Voltage Power Disconnector

Automatic Power-Down Protection Circuit
Line Dropout Detector
Electronic Crowbar
Fast Overvoltage Protector

OVERVOLTAGE PROTECTOR

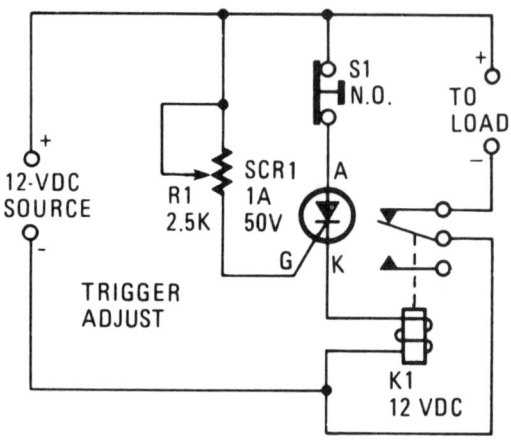

A silicon-controlled rectifier is installed in parallel with the 12-V line and connected to a normally closed 12-V relay, K1. The SCR's gate circuit is used to sample the applied voltage. As long as the applied voltage stays below a given value, SCR1 remains off and K1's contacts remain closed, thereby supplying power to the load. When the source voltage rises above 12 V, sufficient current is applied to the gate of SCR1 to trigger it into conduction. The trigger point of SCR1 depends on the setting of R1. Once SCR1 is triggered (activating the relay), K1's contacts open and halt current flow to the load.

HANDS-ON ELECTRONICS   *Fig. 20-1*

HIGH-SPEED ELECTRONIC CIRCUIT BREAKER

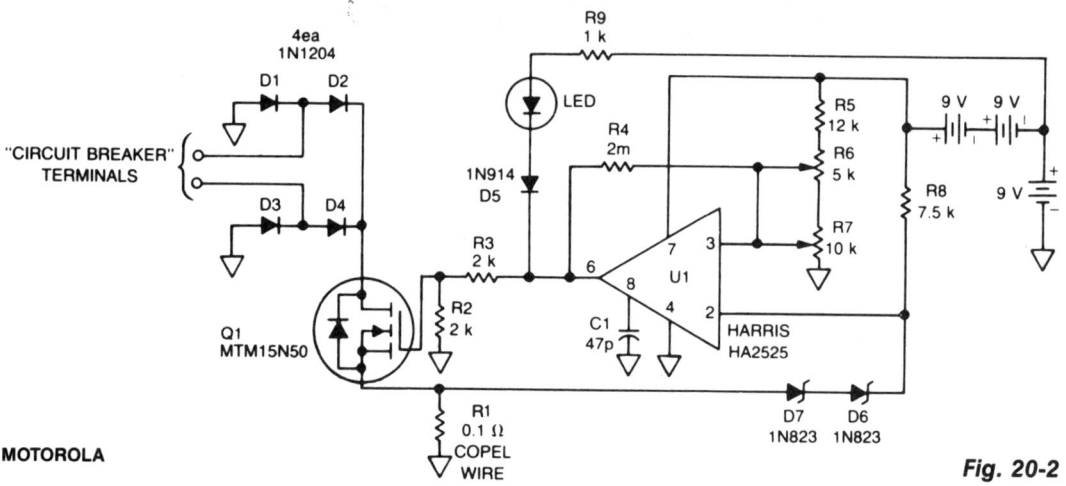

MOTOROLA   *Fig. 20-2*

This 115 Vac, electronic circuit breaker uses the low drive power, low on resistance and fast turn off of the TMOS MTM15N50. The trip point is adjustable, LED fault indication is provided and battery power provides complete circuit isolation.

The two "circuit breaker" terminals are across one leg of a full-wave diode bridge consisting of D1 through D4. Normally, Q1 is turned on so that the circuit breaker looks like a very low resistance. One input to comparator U1 is a fraction of the internal battery voltage and the other input is the drop across zeners D6 and D7 and the voltage drop across R1. If excessive current is drawn, the voltage drop across R1 increases beyond the comparator threshold (determined by the setting of R6), U1 output goes low, Q1 turns off, and the circuit breaker "opens." When this occurs, the LED fault indicator is illuminated.

12-ns CIRCUIT BREAKER

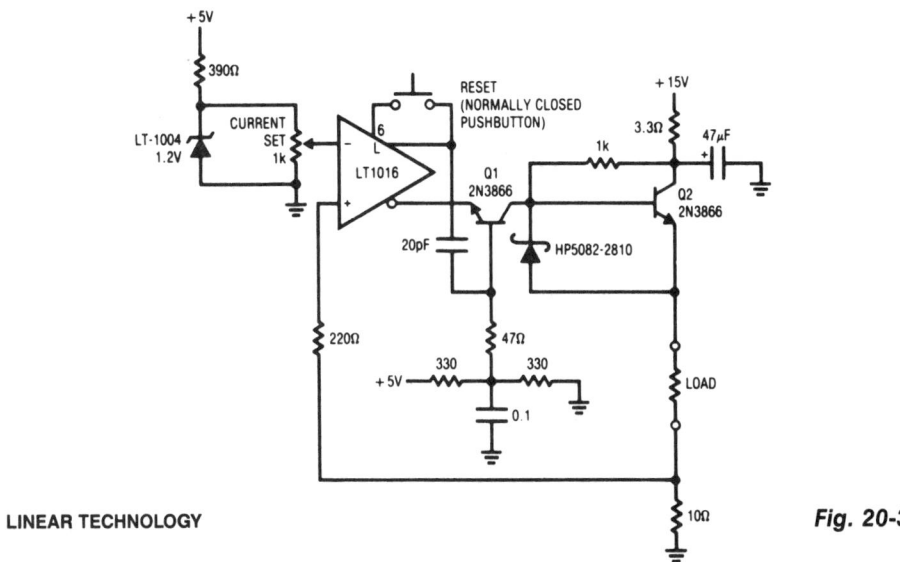

LINEAR TECHNOLOGY

Fig. 20-3

This circuit will turn off current in a load 12 ns after it exceeds a preset value. Under normal conditions, the voltage across the 10-Ω shunt is smaller than the potential at the LT1016's negative input. This keeps Q1 off and Q2 receives bias, driving the load. When an overload occurs, the current through the 10-Ω sense resistor begins to increase. When this current exceeds the preset value, the LT1016's outputs reverse. This provides ideal turn-on drive for Q1 and it cuts off Q2 in 5 ns. The delay from the onset of excessive load current to complete shutdown is just 13 ns. Once the circuit has triggered, the LT1016 is held in its latched state by feedback from the noninverting output. When the load fault has been cleared, the pushbutton can be used to reset the circuit.

LOW-VOLTAGE POWER DISCONNECTOR

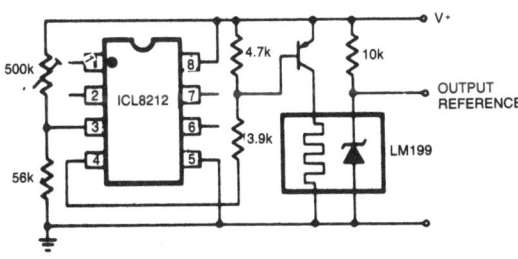

INTERSIL

Fig. 20-4

There are some classes of circuits that require the power supply to be disconnected if the power supply voltage falls below a certain value. As an example, the National LM199 precision reference has an on-chip heater, which malfunctions with supply voltages below 9 V and causes an excessive device temperature. The ICL8212 can be used to detect a power supply voltage of 9 V and turn the power supply off to the LM199 heater section below that voltage.

AUTOMATIC POWER-DOWN PROTECTION CIRCUIT

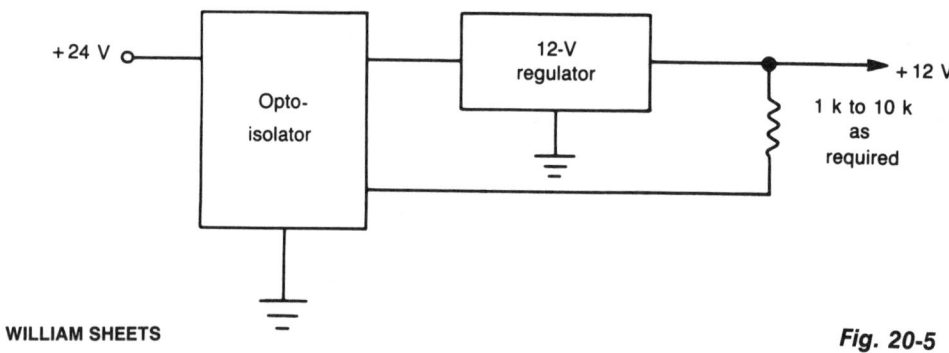

WILLIAM SHEETS

Fig. 20-5

This circuit is faster than a fuse and automatically resets itself when a short is removed. The normal regulated dc input line is opened and the phototransistor of the optoisolator is connected in series with the source and regulator. Between the output of the regulator and ground is an LED and an associated current-limiting resistor, placed physically close to the surface of the photosensitive device. As long as the regulator is delivering its rated output, the LED glows and causes the photo device to have a low resistance. Full current is thus allowed to flow. If a short circuit occurs on the output side of the regulator, the LED goes dark, the resistance of the photo device increases, and the regulator shuts off. When the short is removed, the LED glows, and the regulator resumes operation.

LINE DROPOUT DETECTOR

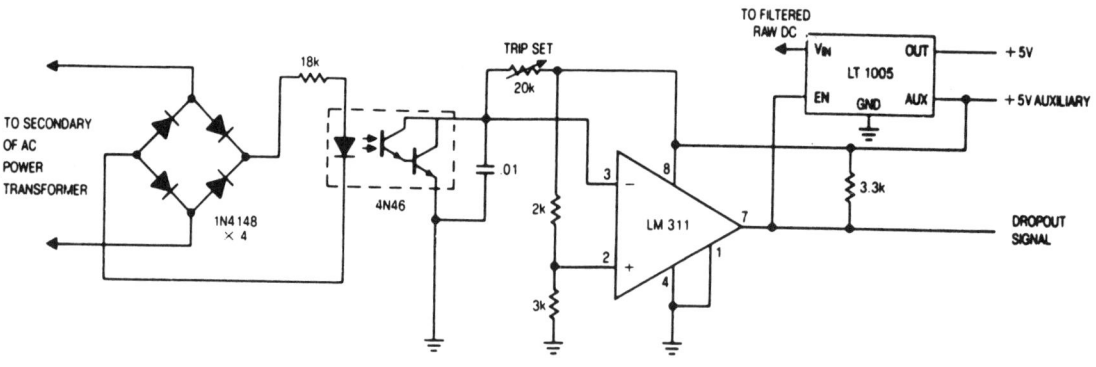

LINEAR TECHNOLOGY

Fig. 20-6

ELECTRONIC CROWBAR

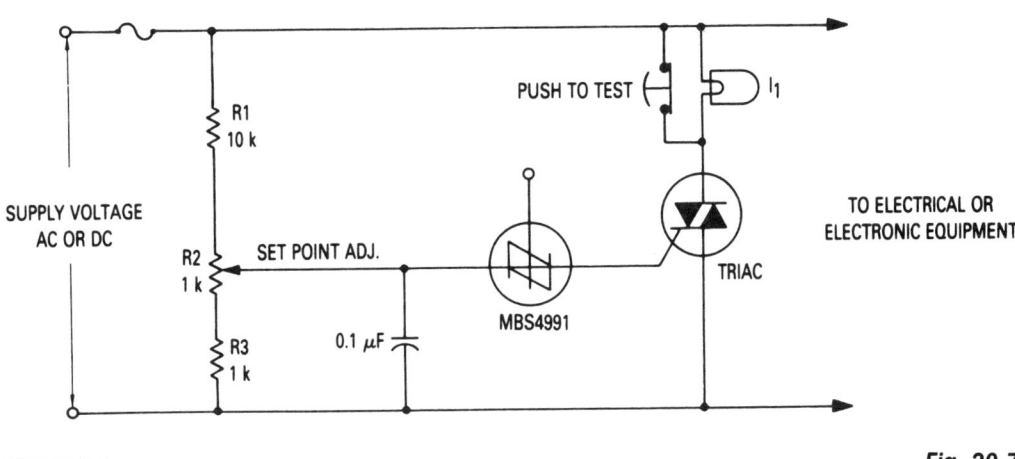

PUSH TO TEST

I_1

R1
10 k

SUPPLY VOLTAGE
AC OR DC

R2
1 k

SET POINT ADJ.

TRIAC

TO ELECTRICAL OR
ELECTRONIC EQUIPMENT

MBS4991

R3
1 k

0.1 μF

MOTOROLA

Fig. 20-7

Where it is desirable to shut down equipment, rather than allow it to operate on excessive supply voltage, an electronic "crowbar" circuit can be used to quickly place a short circuit across the power lines, thereby dropping the voltage across the protected device to near zero and blowing a fuse. Because the triac and SBS are both bilateral devices, the circuit is equally useful on ac or dc supply lines. With the values shown for R1, R2, and R3, the crowbar operating point can be adjusted over the range of 60 to 120 Vdc or 42 to 84 Vac. The resistor values can be changed to cover a different range of supply voltages. The voltage rating of the triac must be greater than the highest operating point, as set by R2. I_1 is a low-power incandescent lamp with a voltage rating that is equal to the supply voltage. It can be used to check the set point and operation of the unit by opening the test switch and adjusting the input or set point to fire the SBS.

FAST OVERVOLTAGE PROTECTOR

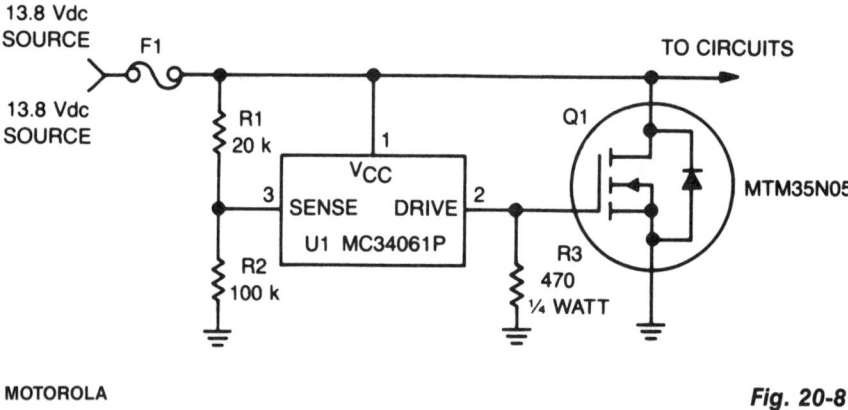

MOTOROLA

Fig. 20-8

This circuit protects expensive portable equipment against all types of improper hookups and environmental hazards that could cause an overvoltage condition. It operates very quickly and does not latchup, that is, it recovers when the overvoltage condition is removed. In contrast, SCR overvoltage circuits can latch and do not recover, unless the power is removed.

Here, U1 senses an overvoltage condition when the drop across R1 exceeds 2.5 V. This causes U1 to apply a positive signal to the gate of Q1, turning it on and shorting the line going to the external circuits. Fuse 1 opens if the transient condition lasts long enough to exceed the i^2t rating.

21

Peak Detectors

The sources of the following circuits are contained in the Sources section, which begins on page 214. The figure number in the box of each circuit correlates to the source entry in the Sources section.

WIDE-RANGE PEAK DETECTOR

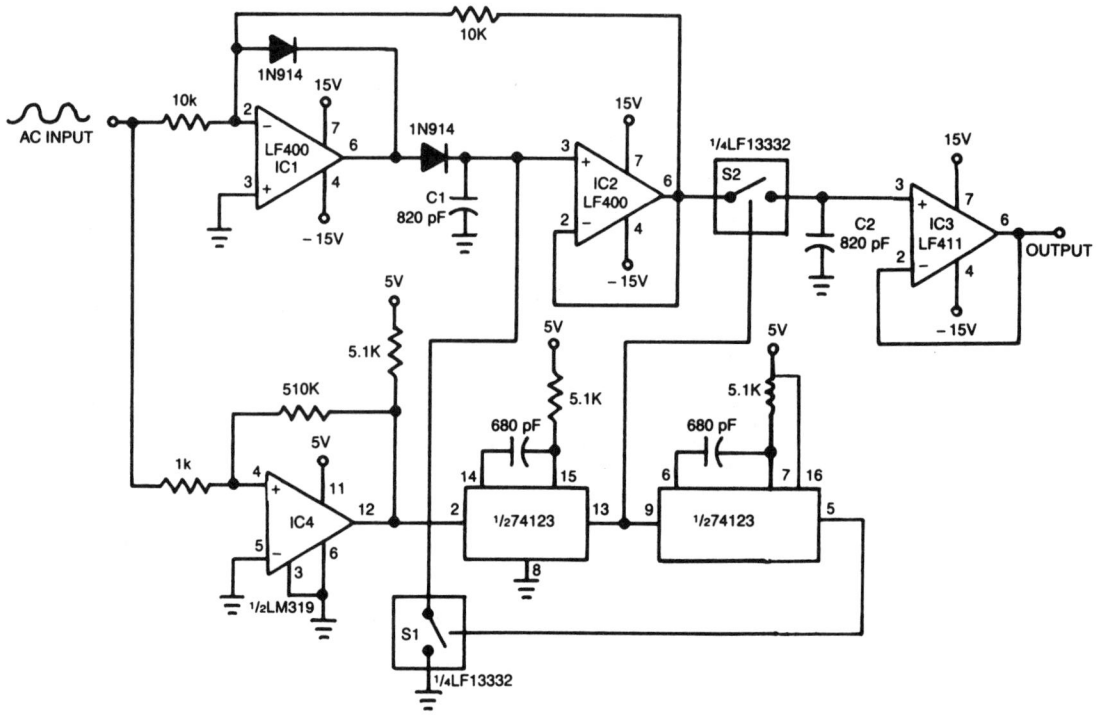

Fig. 21-1

IC1 and IC2 form an inverting half-wave precision-rectifier/peak-detector circuit. Negative input-signal, swings with peaks larger than the voltage on C1, cause this capacitor to charge to the new peak voltage. The capacitor holds this voltage until a larger signal peak arrives. When the input swings high, comparator IC4 detects the zero crossing and triggers the one-shot multivibrator. The one shot closes FET switch S2, and thereby causes C2 to charge to the peak voltage held on C1 during the previous half cycle. The second one shot then produces a pulse that causes FET switch S1 to discharge C1. If the next negative signal-input peak is different from the previous one, the circuit captures it and it appears at IC3's output during the next half cycle. The peak detector thus resets itself once every input-waveform cycle. Notice that the zero crossings are necessary to trigger the switches; therefore, the circuit is usable only with ac signals.

HIGH-FREQUENCY PEAK DETECTOR

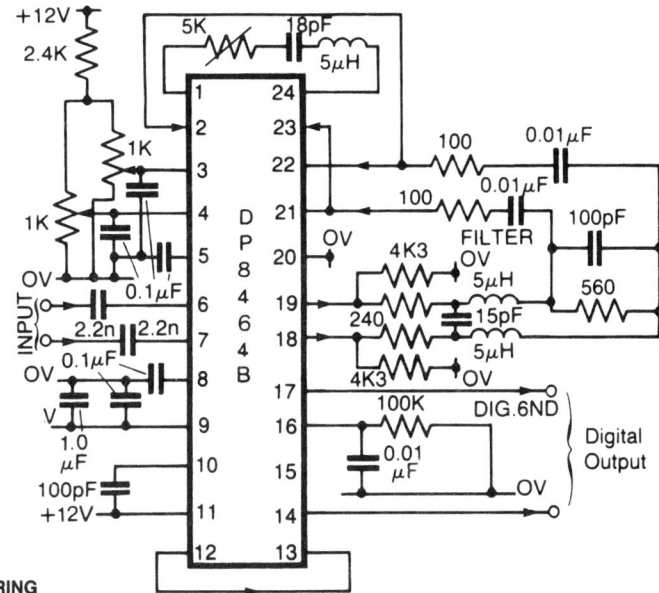

ELECTRONIC ENGINEERING

Fig. 21-2

National Semiconductor's DP8464B is primarily intended for use in disk systems as a pulse detector. However, it can be easily used as a general-purpose peak detector for analog signals up to 5 MHz. The chip can handle signals between 20 and 66 mV peak-to-peak. The circuit includes a filter with constant group-delay characteristics to band limit the signal. Typically, the -3-dB point for this filter will be at about 1.5 times the highest frequency of interest. This differentiator network between pins 1 and 24 can be as simple as a capacitor or it can be more complex to band-limit the differentiator response.

TACHOMETER/SINGLE-PULSE GENERATOR/
POWER-LOSS DETECTOR/PEAK DETECTOR

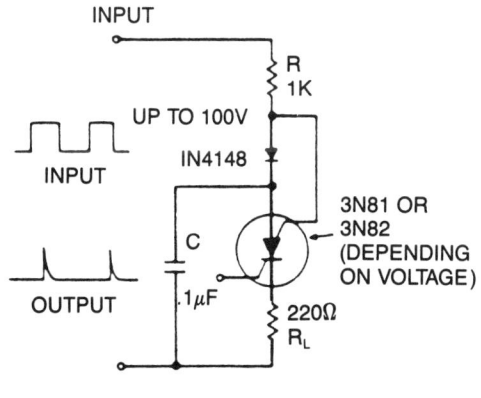

A positive-going input charges C through the IN4148 and R. The diode keeps the SCS off. A negative going input supplies anode-gate current triggering on the SCS discharging C through R_L.

GENERAL ELECTRIC

Fig. 21-3

POSITIVE PEAK DETECTOR I

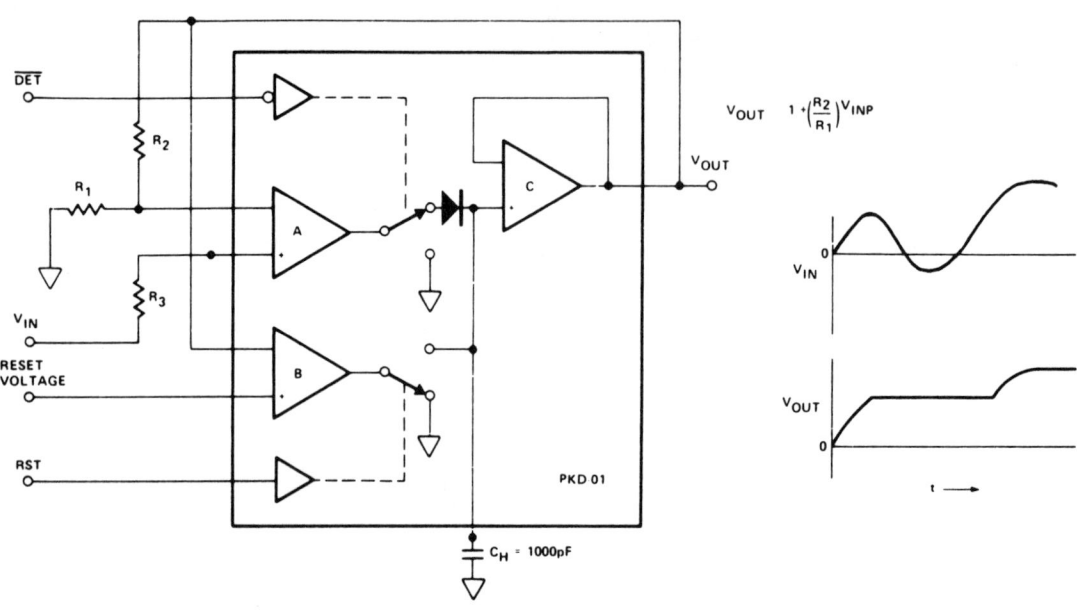

$$V_{OUT} = 1 + \left(\frac{R_2}{R_1}\right) V_{INP}$$

$C_H = 1000pF$

PKD-01

NEGATIVE PEAK DETECTOR I

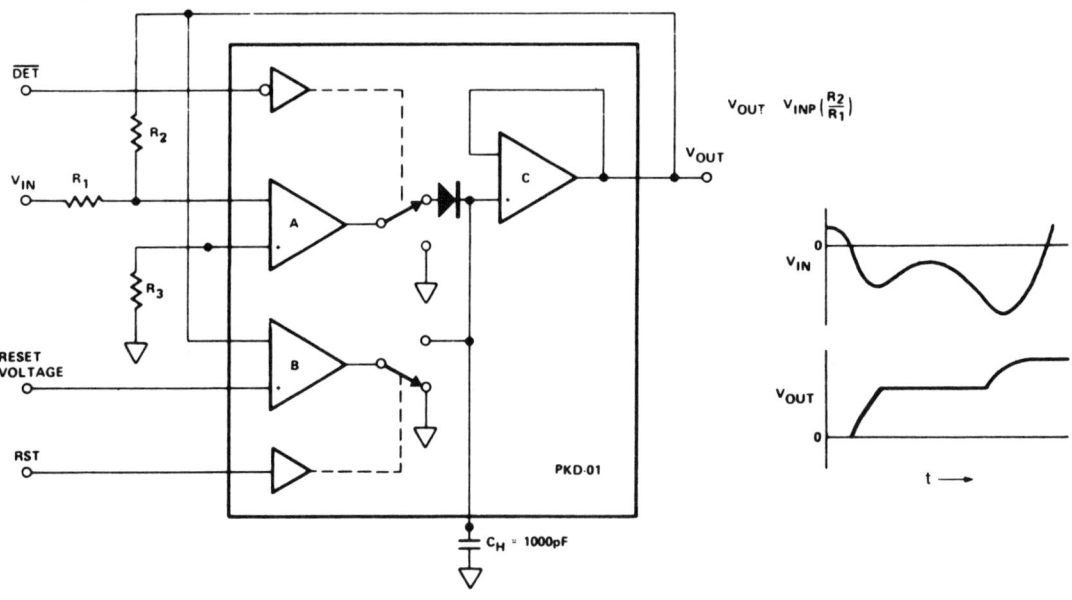

$$V_{OUT} = V_{INP}\left(\frac{R_2}{R_1}\right)$$

$C_H = 1000pF$

PKD-01

POSITIVE PEAK DETECTOR II

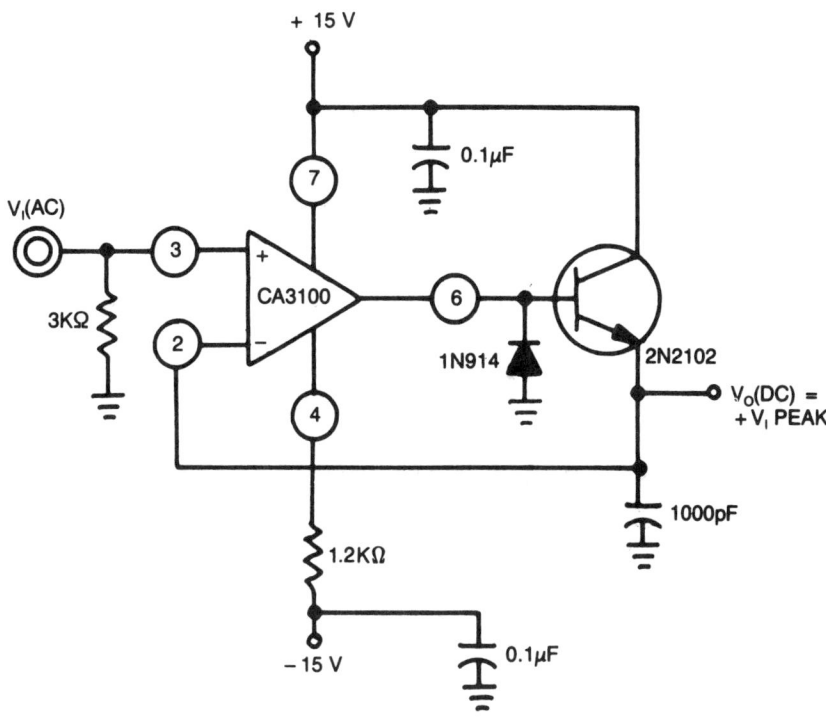

GE/RCA

Fig. 21-6

This peak detector uses a CA3100 BiMOS op amp as a wideband noninverting amplifier to provide essentially constant gain for a wide range of input frequencies. The IN914 clips the negative half of V_{IN} $(R_4)/(R_3)$ (R_5). A 500-μA load current is constant for all load values and the output reflects only positive input peaks.

LOW-DRIFT PEAK DETECTOR

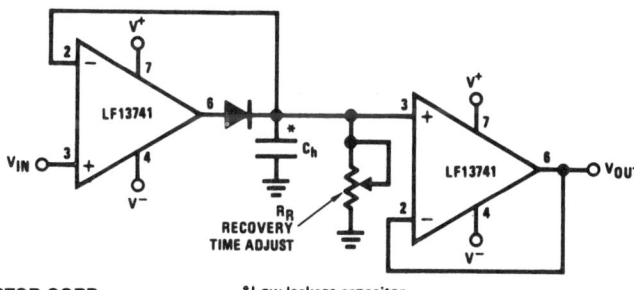

NATIONAL SEMICONDUCTOR CORP.

*Low leakage capacitor

Fig. 21-7

This circuit uses op amp U1 to compensate for the offset in peak detector diode D1. Across C_h is the exact peak voltage; U2 is used as a voltage follower to read this voltage.

DIGITAL PEAK DETECTOR

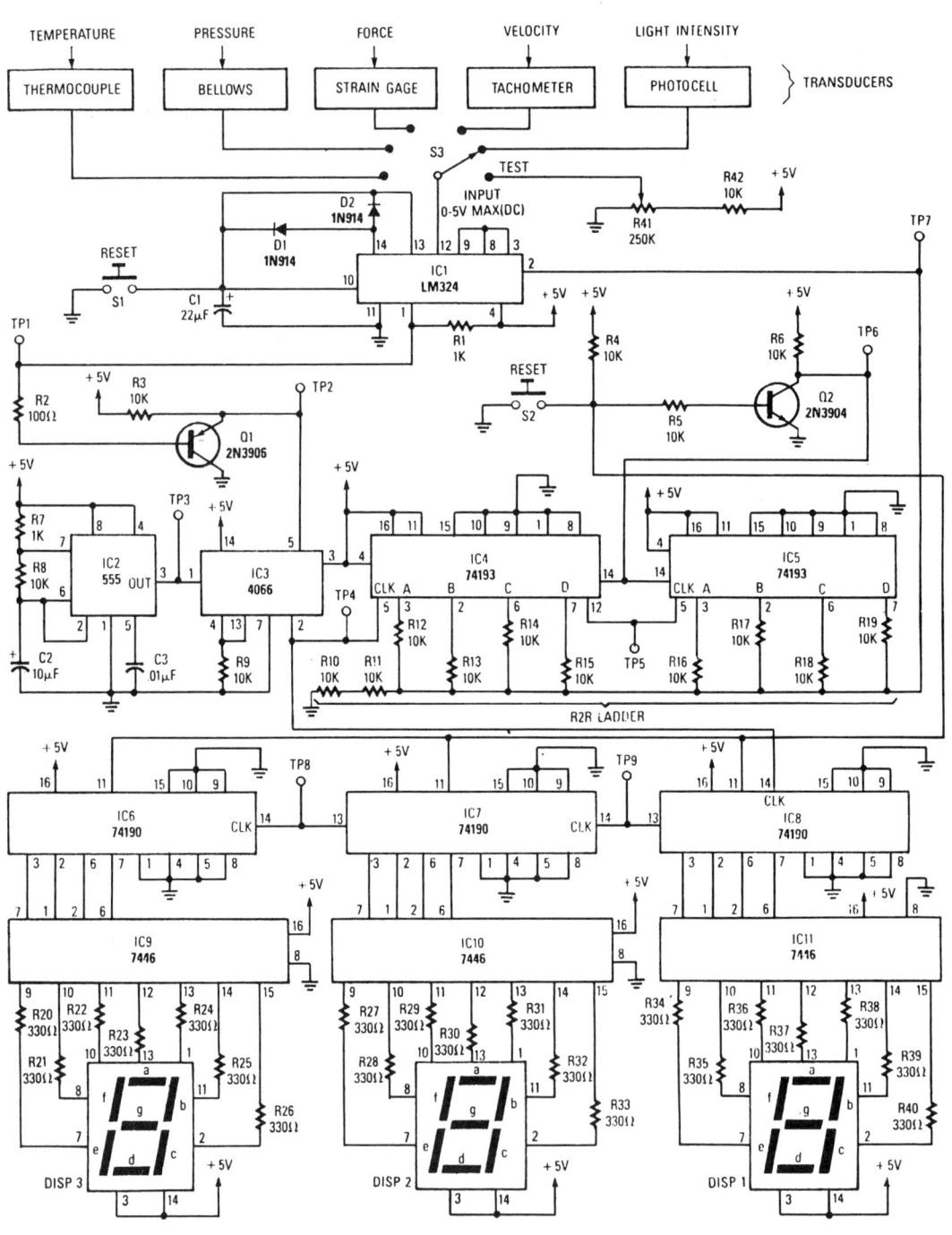

DIGITAL PEAK DETECTOR *(Cont.)*

The peak detector tracks and holds, using the charge-storing ability of a capacitor, the highest output voltage from a transducer. Initially, the voltage on the inverting input of the comparator is at ground level. As a small voltage (0 to 5 V) is captured by the peak detector and presented to the comparator's noninverting input, the output will swing high, which asserts the bilateral switch; clock pulses now pass through the switch to clock both the BCD and binary counters. The outputs of the binary counters are connected to an R2R ladder network, which functions as a digital-to-analog converter. As the binary count increases, the R2R ladder voltage also increases until it reaches a point slightly above the voltage of the peak detector; at that instant, the comparator output swings low, which disables the bilateral switch and stops the counters. The number displayed on the 7-segment LEDs will represent a value equivalent to the transducer's output.

HIGH-BANDWIDTH PEAK DETECTOR

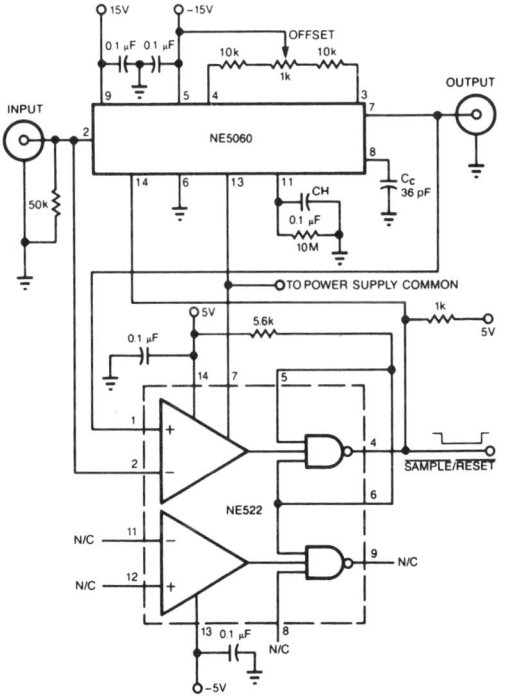

The high-speed peak detector uses a highly accurate, fast s/h amplifier controlled by a high-speed comparator. The s/h amplifier holds the peak voltage, until the comparator switches the amp to its sample mode, to capture a new, higher voltage level. The circuit handles all common wave shapes and exhibits 5% accuracy from 50 Hz to 2 MHz.

The comparator's output decreases when the input signal exceeds the value of the currently held output. This transition puts the s/h amplifier into sample mode. Once the output reaches the value of the input, or the input signal falls below the output's level, the comparator's output increases; the high output brings the s/h amplifier back to the hold mode, thereby holding the peak value of the input signal. Reset the circuit by lowering the value of pin 4 of the NE522 comparator, which in turn allows the NE5060 s/h amplifier to acquire the input. The NE522 comparator has an open-collector output.

EDN

Fig. 21-9

POSITIVE PEAK DETECTOR III

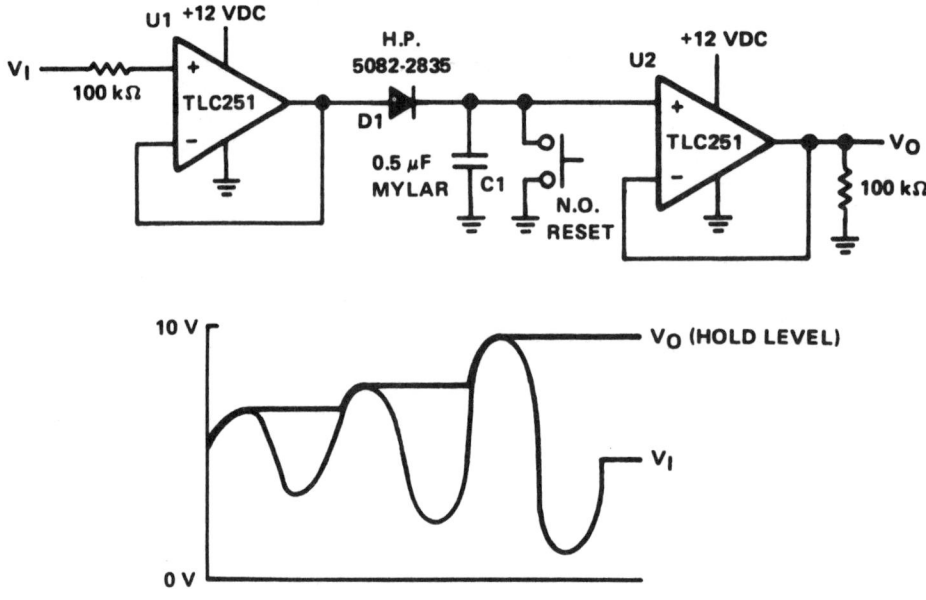

Fig. 21-10

The purpose of the circuit is to hold the peak of the input voltage on capacitor C1, and read the value, V_O, at the output of U2. Op amps U1 and U2 are connected as voltage followers. When a signal is applied to V_I, C1 will charge to this same voltage through diode D1. This positive peak voltage on C1 will maintain V_O at this level until the capacitor is reset (shorted). Of course, higher positive peaks will raise this level and lower power peaks will be ignored. C1 can be reset manually with a switch, or electronically with an FET that is normally off. The capacitor specified for C_1 should have low leakage and low dielectric absorption. Diode D1 should have low leakage. Peak values of negative-polarity signals can be detected by reversing D1.

ULTRA-LOW-DRIFT PEAK DETECTOR

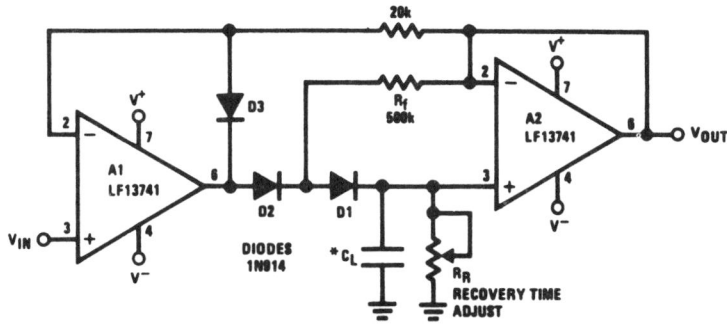

- By adding D1 and R_f, $V_{D1} = 0$ during hold mode. Leakage of D2 provided by feedback path through R_f.
- Leakage of circuit is I_B plus leakage of C_h.
- D3 clamps V_{OUT} A1 to $V_{IN} - V_{D3}$ to improve speed and to limit the reverse bias of D2.
- Maximum input frequency should be $<< 1/2\pi R_f C_{D2}$, where C_{D2} is the shunt capacitance of D2.

*Low leakage capacitor

NATIONAL SEMICONDUCTOR

Fig. 21-11

EDGE DETECTOR

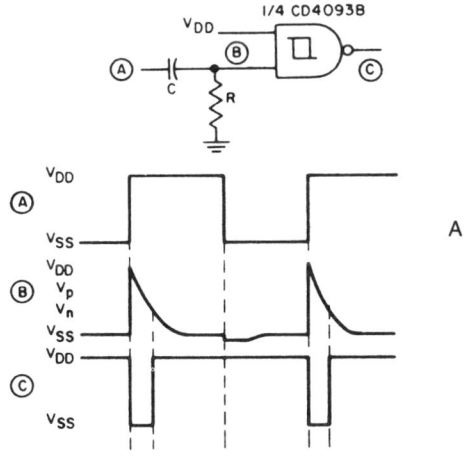

This circuit provides a short negative-going output pulse for every positive-going edge at the input. The input waveform is coupled to the input by capacitor C; the pulse length depends, as before, on R and C. If a negative-going edge detector is required, the circuit in B should be used.

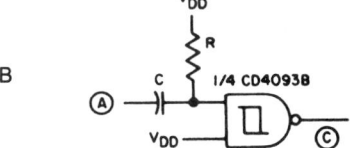

RCA

Fig. 21-12

PEAK DETECTOR I

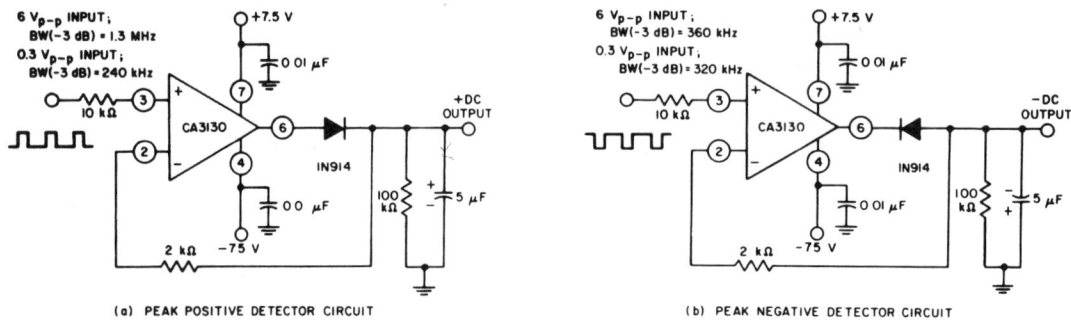

(a) PEAK POSITIVE DETECTOR CIRCUIT

(b) PEAK NEGATIVE DETECTOR CIRCUIT

GENERAL ELECTRIC/RCA

Fig. 21-13

Circuits are easily implemented using the CA3130 BiMOS op amp. For large-signal inputs, the bandwidth of the peak-negative circuit is less than that of the peak-positive circuit. The second stage of the CA3130 limits bandwidth in this case.

HIGH-SPEED PEAK DETECTOR

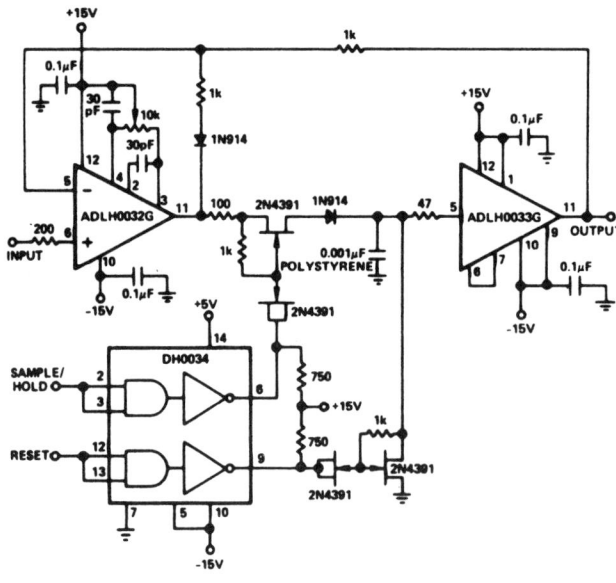

ANALOG DEVICES

Fig. 21-14

WIDE-BANDWIDTH PEAK DETECTOR

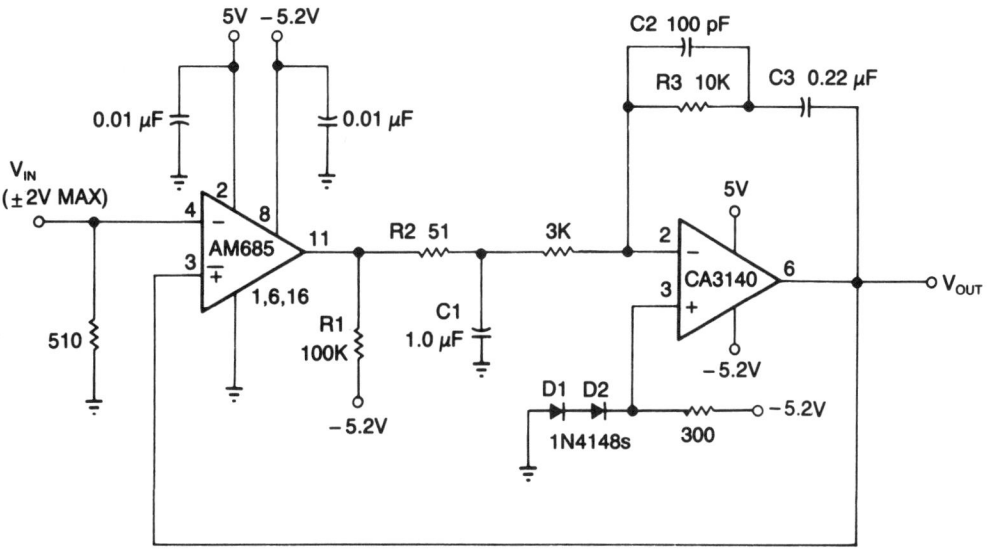

EDN

Fig. 21-15

This circuit can detect the positive peaks for signal frequencies higher than 5 MHz. It yields ±1% accuracy for 400 mV to 4 V pk-pk signal amplitudes on sine, square, and triangular waveforms. The AM685 comparator output increases whenever V_{IN} is a greater negative voltage than V_{OUT}; the high comparator output, in turn, charges C1 in a positive direction. The CA3140 op amp amplifies the C1 voltage with respect to the ECL-switching-threshold voltage (-1.3 V) developed by diodes D1 and D2. For repetitive waveforms, each cycle boosts V_{OUT} until it equals the peak input value. The peak-detection process is aided by the comparator's open-emitter output, which allows C1 to charge rapidly through R2, but to discharge slowly through R2 and R1. Reducing the value of C1 shortens system-response times. Although the circuit can't detect negative-going peaks, it can be modified to measure the pk-pk value of bipolar signals that are symmetric about ground. To do so, divide V_{OUT} by 2 using two 1-kΩ resistors and feed the comparator $V_{OUT}/2$, rather than V_{OUT}.

ANALOG PEAK DETECTOR WITH DIGITAL HOLD

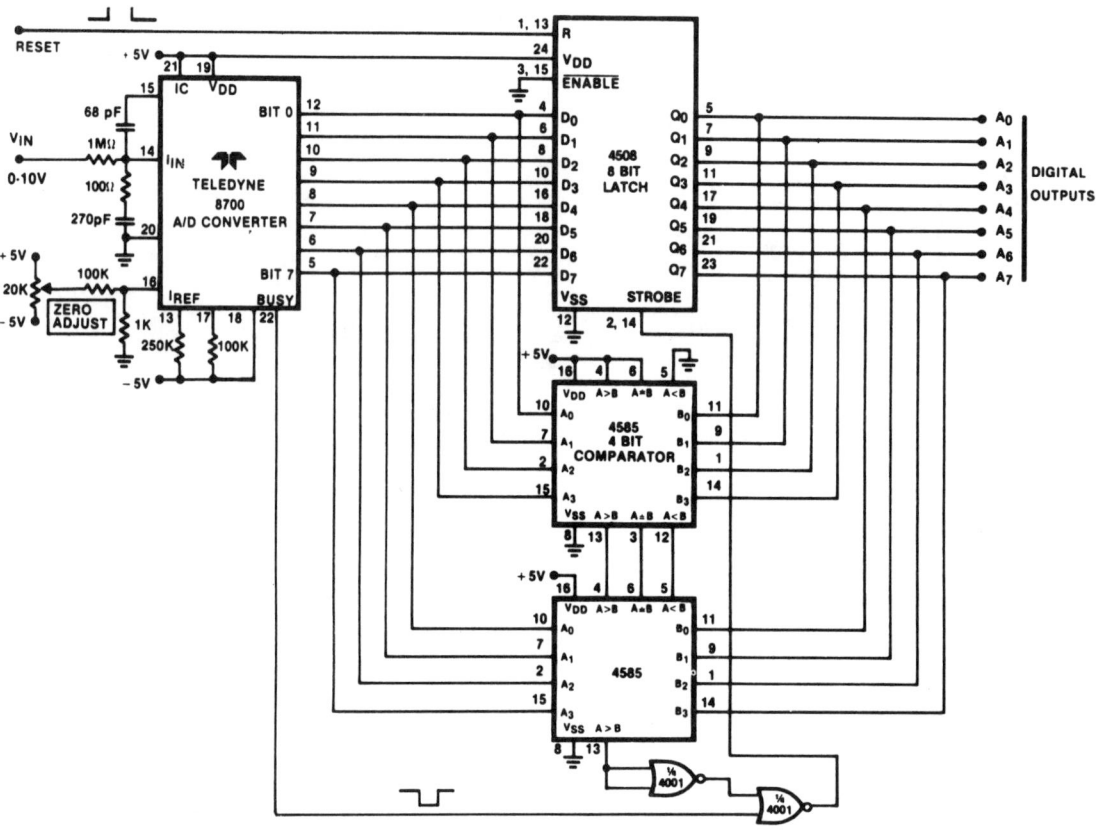

TELEDYNE INDUSTRIES INC.

Fig. 21-16

Analog peak detection is accomplished by repeatedly measuring the input signal with an a/d converter and comparing the current reading with the previous reading. If the current reading is larger than the previous, the current reading is stored in the latch and becomes the new peak value. Because the peak is stored in a CMOS latch, the peak can be stored indefinitely.

POSITIVE PEAK DETECTOR IV

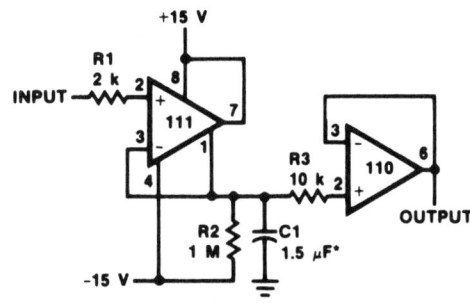

* Solid tantalum

FAIRCHILD CAMERA & INSTRUMENT   *Fig. 21-17*

PRECISION PEAK VOLTAGE DETECTOR WITH LONG MEMORY TIME

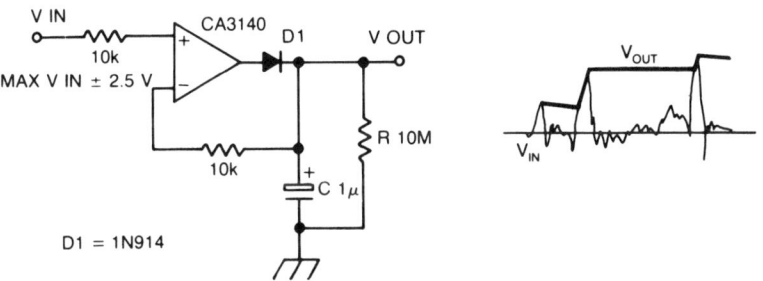

ELECTRONICS TODAY INTERNATIONAL   *Fig. 21-18*

The circuit has negative feedback only for positive signals. The inverting input can only get some feedback when diode D1 is forward biased and only occurs when the input is positive. With a positive input signal, the output of the op amp rises until the inverting input signal reaches the same potential. In so doing, the capacitor (C) is also charged to this potential. When the input goes negative, the diode (D1) becomes reverse biased and the voltage on the capacitor remains, being slowly discharged by the op amp input bias current of 10 pA. Thus, the discharge of the capacitor is dominantly controlled by the resistor (R), giving a time constant of 10 seconds. Thus, the circuit detects the most positive peak voltage and remembers it.

PEAK DETECTOR II

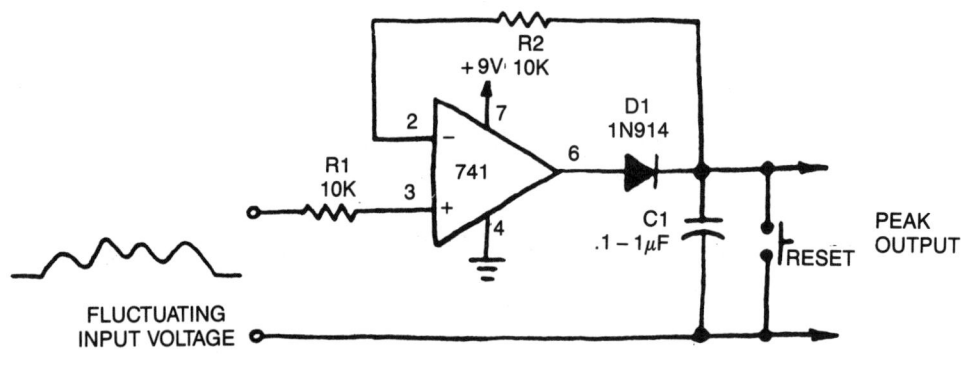

POPULAR ELECTRONICS

Fig. 21-19

The comparator will charge C1 until the voltage across the capacitor equals the input voltage. If subsequent input voltage exceeds that stored in C1, the comparator voltage will go high and charge C1 to a new, higher peak voltage.

NEGATIVE PEAK DETECTOR II

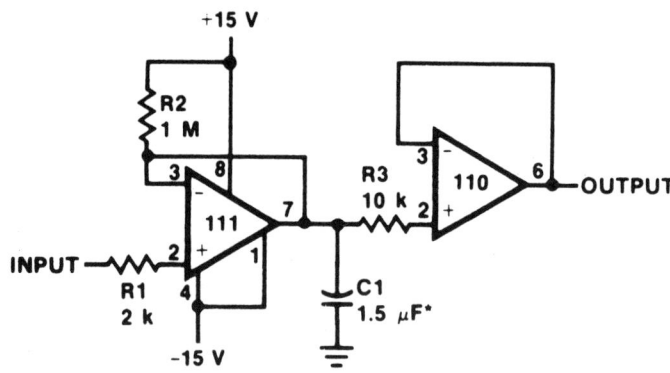

*Solid tantalum

FAIRCHILD CAMERA & INSTRUMENT

Fig. 21-20

22

Phase Detectors

The sources of the following circuits are contained in the Sources section, which begins on page 214. The figure number in the box of each circuit correlates to the source entry in the Sources section.

RC CIRCUIT DETECTS PHASE-SEQUENCE REVERSAL

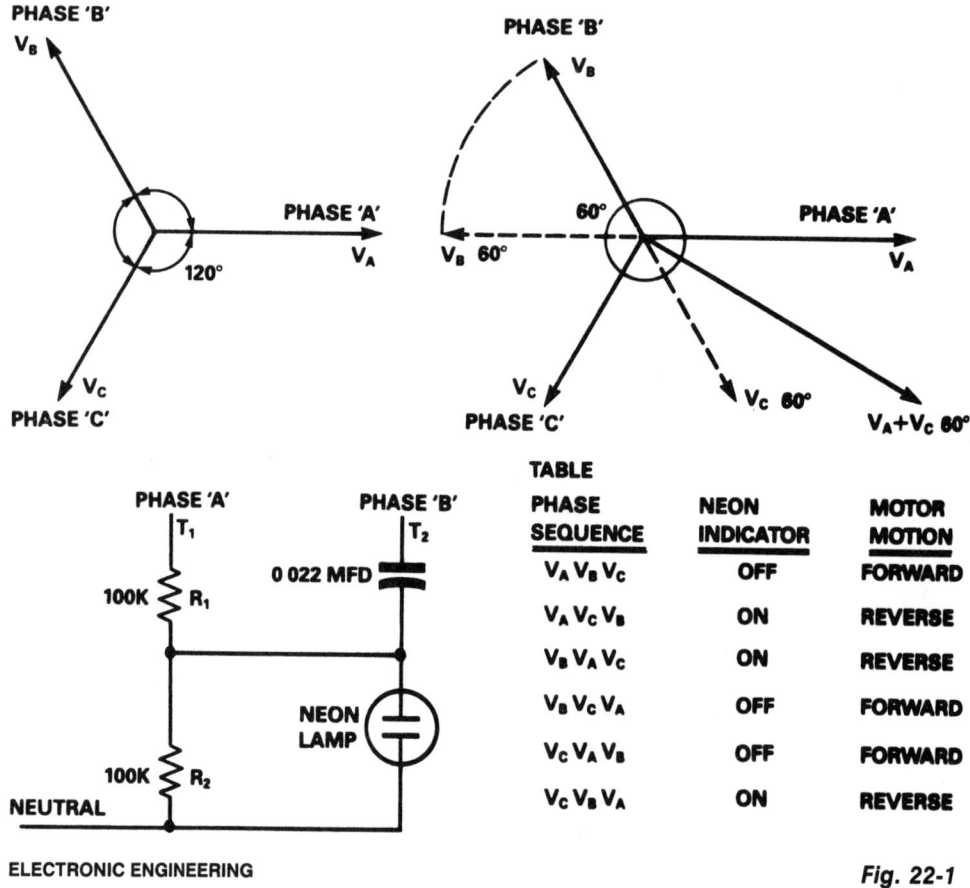

ELECTRONIC ENGINEERING

Fig. 22-1

TABLE

PHASE SEQUENCE	NEON INDICATOR	MOTOR MOTION
V_A V_B V_C	OFF	FORWARD
V_A V_C V_B	ON	REVERSE
V_B V_A V_C	ON	REVERSE
V_B V_C V_A	OFF	FORWARD
V_C V_A V_B	OFF	FORWARD
V_C V_B V_A	ON	REVERSE

Assume the correct phase sequence to be $V_A - V_B - V_C$. The circuit terminals are connected so that T1 gets connected to phase A and T2 to phase B. The capacitor advances the voltage developed across R2 as a result of phase "B" by $\sim 60°$, while the voltage developed across it by phase "A" is in phase with V_A (as shown). The net voltage developed across R2 $\sim$ zero, the neon lamp is not energized, and thereby signals correct phase sequence. If terminal T2 gets connected to phase C, a large voltage, $K(V_A + V_C \ 60°)$, gets developed across R2, energizing the neon indicator to signal reverse phase sequence.

The motor terminals can be connected to the three phases in six different combinations. A three-phase motor will run in the forward direction for three such combinations. For the other three, it will operate in the reverse direction. As shown in the table, the circuit detects all three reverse combinations. This circuit can be wired into any existing motor starter, where the operator can see whether the phase sequence has been altered, before starting the machine.

PHASE INDICATOR

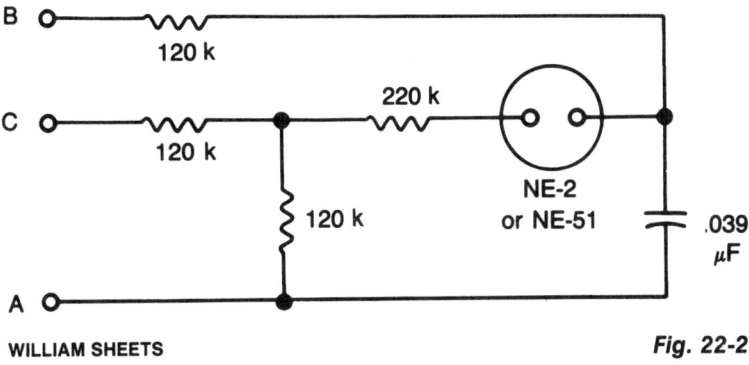

WILLIAM SHEETS

Fig. 22-2

The circuit provides a simple means of determining the succession of phases of a 3-phase 120-V source used in synchro work. Terminals A, B, and C are connected to the three terminals of the source to be checked. If the neon bulb lights, interchange any two leads; the light then extinguishes and A, B, and C indicate the correct sequence. If power on any one line is lost, the neon bulb will light. This feature is useful for monitoring purposes.

PHASE-SEQUENCE DETECTOR I

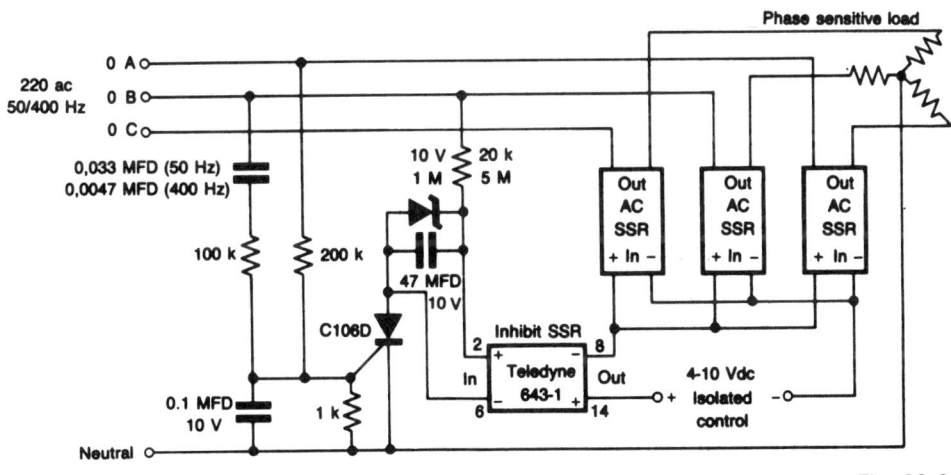

ELECTRONIC ENGINEERING

Fig. 22-3

This circuit prevents damage to the load as a result of incorrect phasing. The three power SSRs are only permitted to turn on for a phase sequence of phase A leading phase B. If phase A lags phase B, the input currents will cancel, and cause the SCR and the inhibit SSR to remain off until the sequence is reversed. The inhibit SSR is included to maintain isolation at the input.

THREE-PHASE TESTER

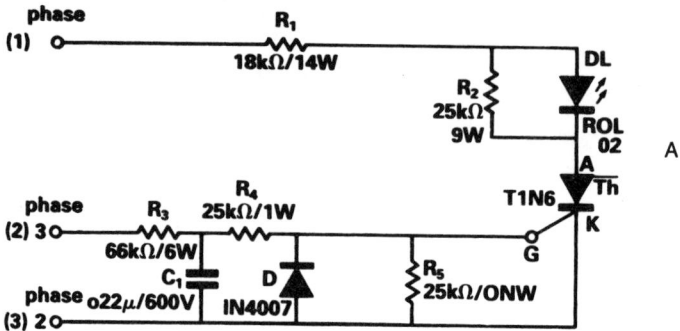

A

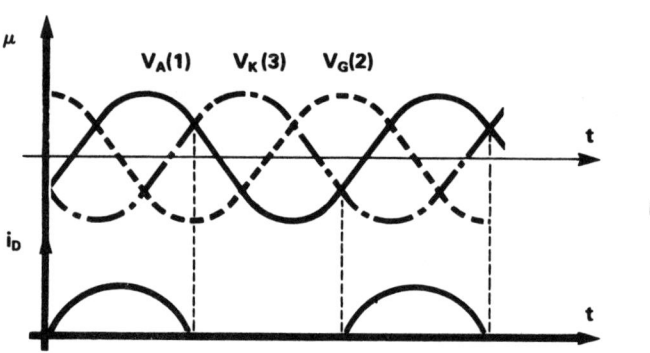

B

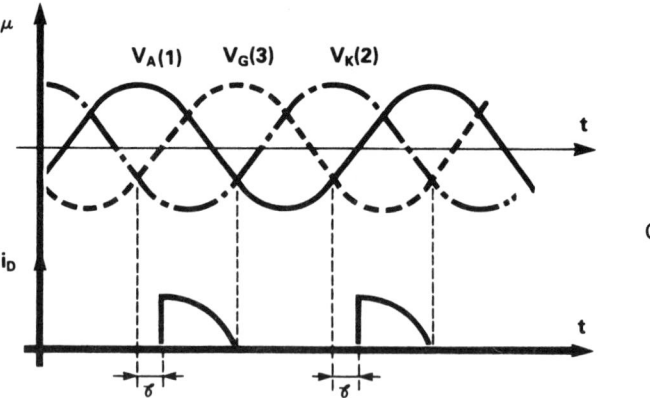

C

Fig. 22-4

THREE-PHASE TESTER *(Cont.)*

This simple three-phase tester uses only a small current thyristor as a main element for testing the right or wrong succession of the three phases, and there is no need for a supplementary power supply.

The basic circuit is shown in Fig. 22-4A. When connecting to the thyristor anode, grid, and cathode, the three phases of the supply network in the sequence phase 1, phase 3, phase 2, are considered as correct; the mean value of the current through the thyristor is relatively high (it is turned on during an entire half-period of one phase). The result is that the LED will emit a normal light.

The wave shapes for the three voltages and the current through the LED for this situation are shown in Fig. 22-4B.

If the three phases are not correctly connected—phase 1 to the anode, phase 2 to the grid, and phase 3 to the cathode, for instance—the thyristor will be turned on for a very short time and the LED will produce a very poor light. The wave shapes for this case are shown in Fig. 22-4C. The delay time is given by the R3/R1/R4 group.

When any of the three phases is missing, there is no current through the thyristor and the LED will emit no light.

PHASE-SEQUENCE DETECTOR II

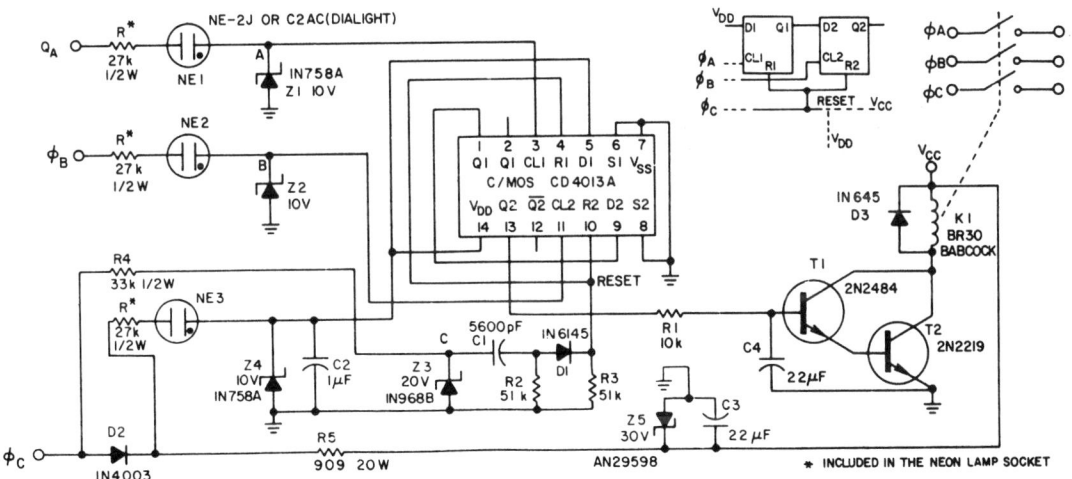

ELECTRONIC DESIGN

Fig. 22-5

This circuit derives its supply voltage, V_{cc} and V_{dd} from ϕ_c. This factor, together with the neon lamps and zener diodes in the phase inputs, establishes the 50% threshold that detects low voltage or absence of one or more phases. Relay K1 energizes for correct phase volts.

SIMPLE PHASE-DETECTOR CIRCUIT

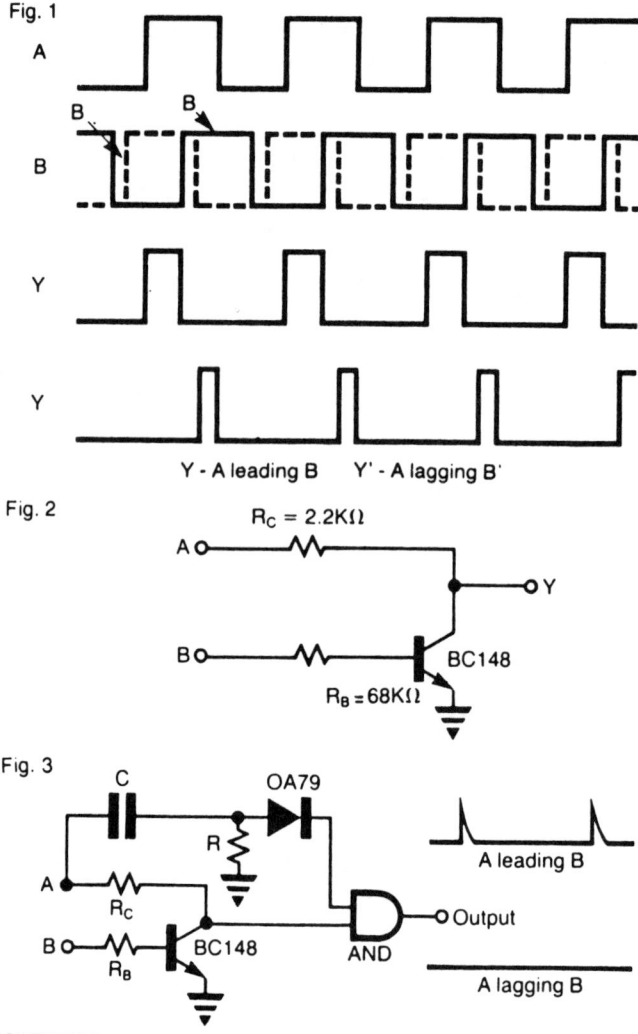

Fig. 1

A

B

Y

Y

Y - A leading B   Y' - A lagging B'

Fig. 2

$R_C = 2.2K\Omega$

A

B

Y

BC148

$R_B = 68K\Omega$

Fig. 3

C   OA79

R

A leading B

A

R_C

B

BC148   AND

R_B

Output

A lagging B

ELECTRONIC ENGINEERING

Fig. 22-6

The operation of the circuit is like an enabled inverter, that is, the output $Y = B$ provided A is high. If A is low, output is low (independent of the state of B). When the signals A and B or B_1 are connected to the inputs A and B of this gate, the output Y is a pulse-train signal (shown as Y or Y_1), which has a pulse duration equal to the phase difference between the two signals. The circuit is directly suitable for phase difference measurement from 0 to 180°. This performance is similar to the circuits like the Exclusive OR gate used for this purpose. With this method leading and lagging positions of the signals can also be found using an AND gate. Phase difference measured along with the leading and lagging information gives complete information about the phases of the two signals between 0 and 360°.

23

Power-Failure Detectors

The sources of the following circuits are contained in the Sources section, which begins on page 214. The figure number in the box of each circuit correlates to the source entry in the Sources section.

POWER-FAILURE DETECTOR

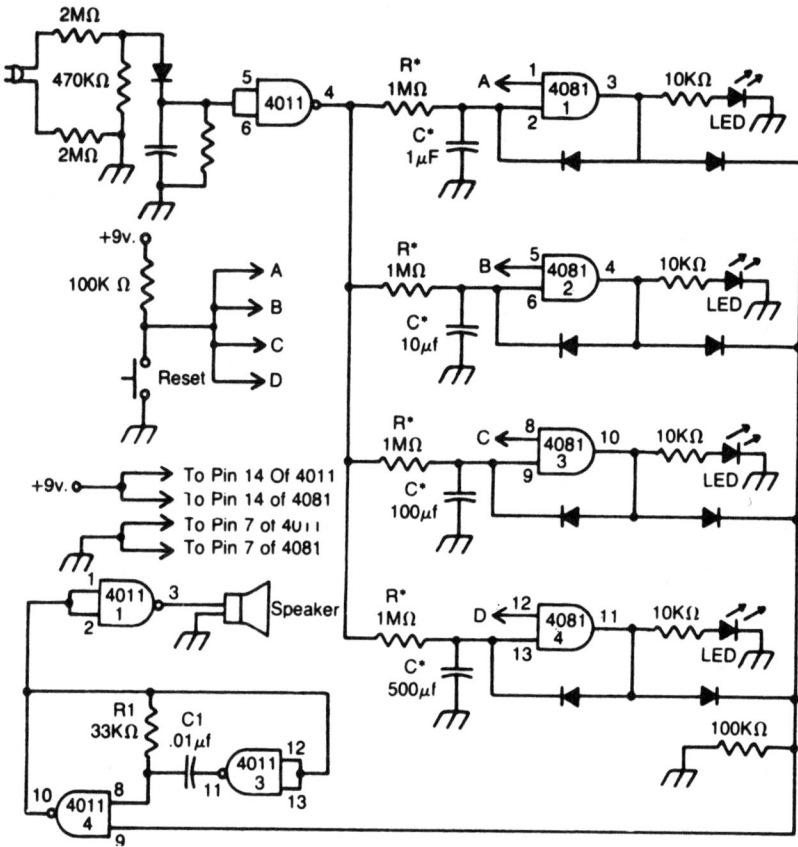

MODERN ELECTRONICS

Fig. 23-1

This circuit indicates that a power outage occurred for 1, 10, 100, and 500 seconds with the values given for R^* and C^*. After a power failure, the circuit can be reset by pushing the reset button.

POWER-FAILURE ALARM I

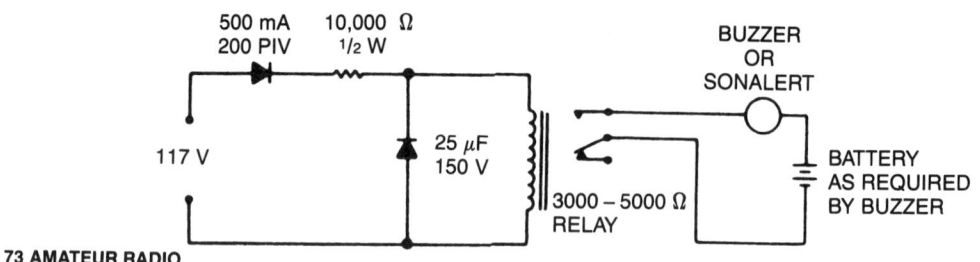

73 AMATEUR RADIO

Fig. 23-2

When the power is on, the relay is held open. When the power fails, the buzzer-circuit contacts close.

LINE-VOLTAGE MONITOR

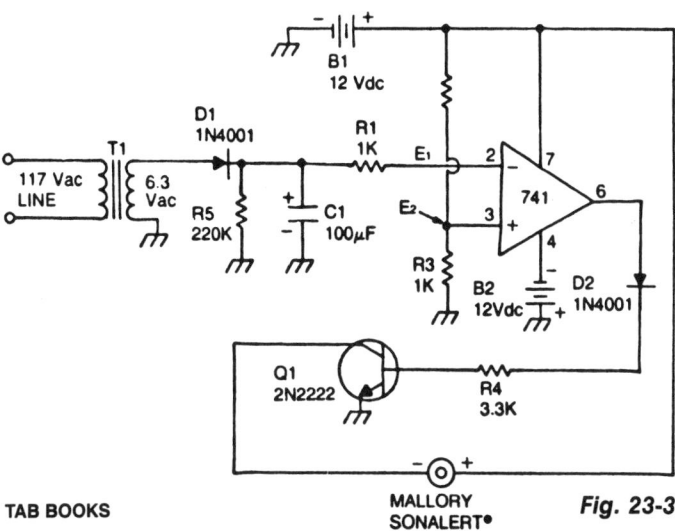

TAB BOOKS

MALLORY
SONALERT•

Fig. 23-3

This circuit uses a type 741 op amp as a voltage comparator. One input of the 741 is connected to a reference voltage (a 12-V battery) through a resistor voltage divider. The potential at the noninverting input of the 741 is approximately 3 V. The inverting input of the op amp comparator is connected to the output of a line-operated 8-V power supply. When the ac power main fails, T1 will no longer be energized, so the charge stored in capacitor C1 will begin to discharge through resistor R5. When voltage on the capacitor drops below the reference voltage of 3 V, the output of the comparator becomes high. This output condition will forward-bias transistor Q1, causing the Sonalert to sound the alarm. The time constant of the R_5/C_1 combination is 22 seconds—long enough to prevent noise from triggering the alarm.

POWER-FAILURE ALARM II

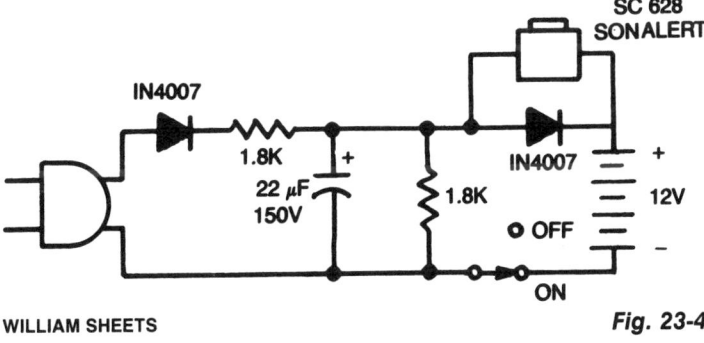

WILLIAM SHEETS

Fig. 23-4

With power ac off, the alarm sounds when S1 is closed on. The 12-V battery is kept charged when the circuit is plugged in and the switch is left on.

POWER-FAILURE ALARM III

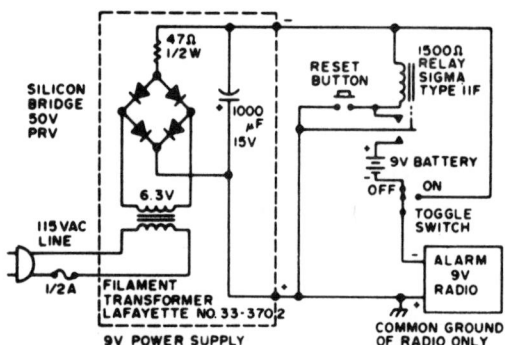

SILICON BRIDGE 50V PRV

47Ω 1/2 W

1000 µF 15V

RESET BUTTON

1500Ω RELAY SIGMA TYPE 11F

9V BATTERY

OFF ON

TOGGLE SWITCH

6.3V

115 VAC LINE

1/2 A

FILAMENT TRANSFORMER LAFAYETTE NO. 33-370.2

9V POWER SUPPLY

ALARM 9V RADIO

COMMON GROUND OF RADIO ONLY

If the power fails, the radio alarm goes on—no loud siren, bell, or whistle. Even if the power is restored, the alarm stays on until the reset button is pushed.

73 AMATEUR RADIO

Fig. 23-5

24

Proximity Detectors

The sources of the following circuits are contained in the Sources section, which begins on page 214. The figure number in the box of each circuit correlates to the source entry in the Sources section.

Proximity Alarm
Proximity Switch
SCR Proximity Alarm
UHF Movement Detector
Capacitive Sensor Alarm

Capacitance Relay
Self-Biased Proximity Sensor
Touch Switch or Proximity Detector
Field-Disturbance Sensor/Alarm
Proximity Detector

Proximity Sensor

PROXIMITY ALARM

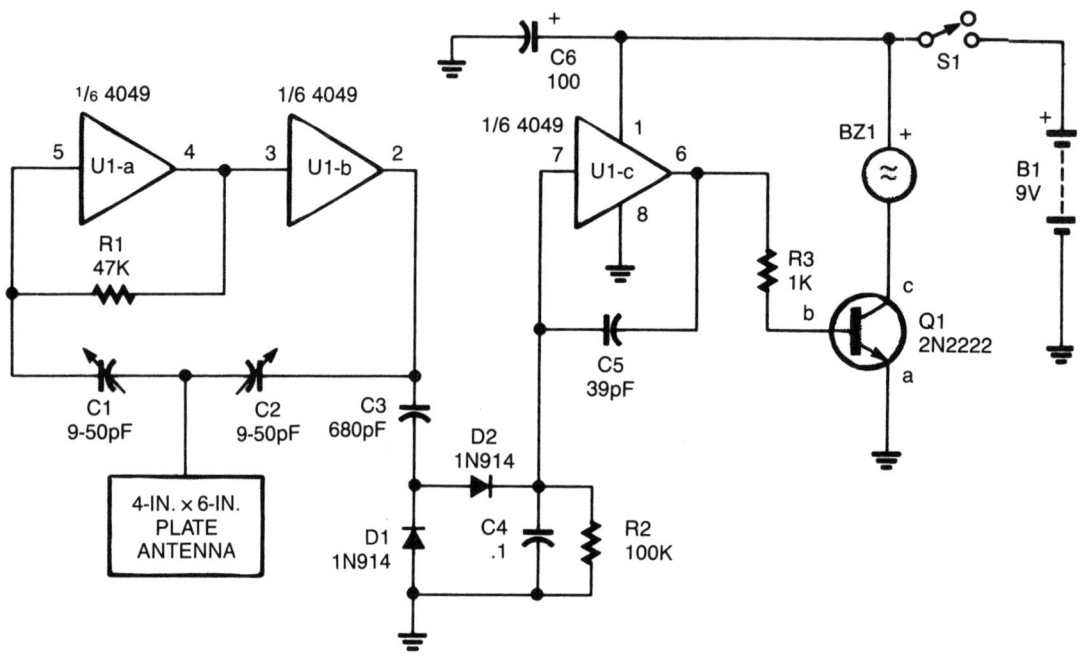

Fig. 24-1

Inverters U1a and U1b are connected in a simple RC oscillator circuit. The frequency is determined by the values of R_1, C_1, C_2, and the internal characteristics of the integrated circuit. As long as the circuit is oscillating, a positive dc voltage is developed at the output of the voltage-coupler circuit: C3, D1, D2, and C4. The dc voltage is applied to the input of U1c—the third inverter amplifier—keeping its output in a low state, which keeps Q1 turned off so that no sound is produced by BZ1. With C1 and C2 adjusted to the most sensitive point, the pickup plate will detect a hand 3 to 5 inches away and sound an alert. Set C1 and C2 to approximately one-half of their maximum value and apply power to the circuit. The circuit should oscillate and no sound should be audible. Using a nonmetallic screwdriver, carefully adjust C1 and C2 one at a time, to a lower value until the circuit just ceases oscillation: buzzer BZ1 should sound off. Slightly back off either C1 or C2, until the oscillator starts up again; that is the most sensitive setting of the circuit.

PROXIMITY SWITCH

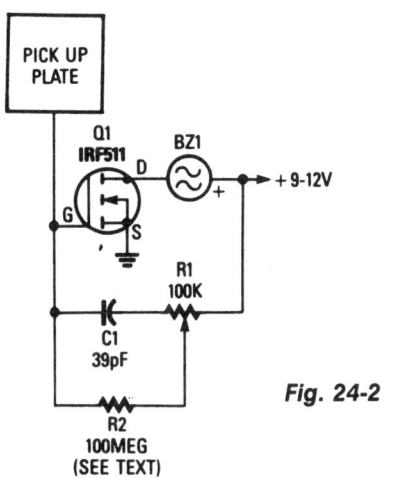

A 3-×-3-inch piece of circuit board, or similar size metal object which functions as the pick-up sensor, is connected to the gate of Q1. A 100-MΩ resistor, R2, isolates Q1's gate from R1, allowing the input impedance to remain very high. If a 100-MΩ resistor cannot be located, just tie five 22-MΩ resistors in series and use that combination for R2. In fact, R2 can be made even higher in value for added sensitivity.

Potentiometer R1 is adjusted to where the piezo buzzer just begins to sound off and then carefully backs off to where the sound ceases. Experimenting with the setting of R1 will help in obtaining the best sensitivity adjustment for the circuit. Resistor R1 can be set to where the pick-up must be contacted to set off the alarm sounder. A relay or other current-hungry component can take the place of the piezo sounder to control most any external circuit.

Fig. 24-2

SCR PROXIMITY ALARM

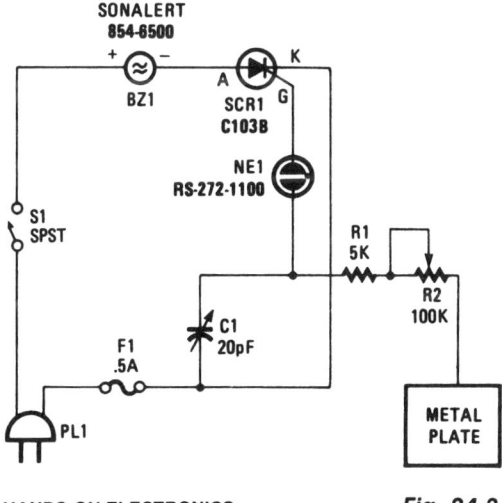

Fig. 24-3

UHF MOVEMENT DETECTOR

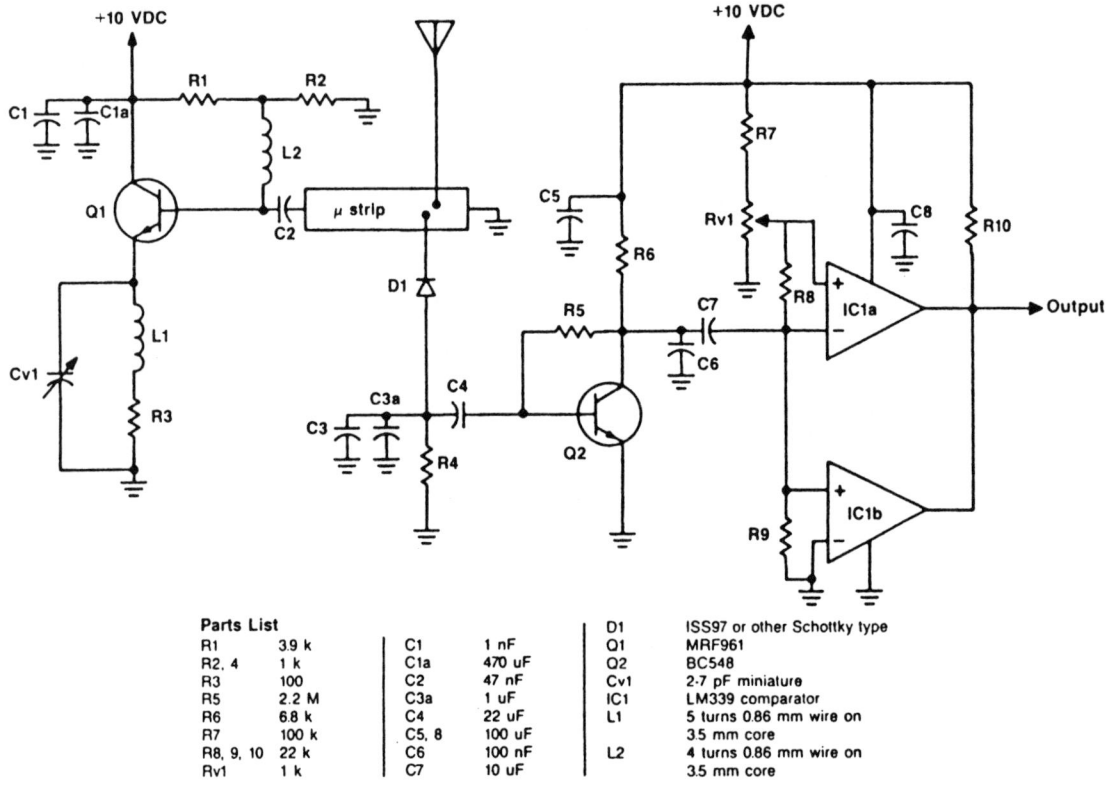

Parts List							
R1	3.9 k	C1	1 nF	D1	ISS97 or other Schottky type		
R2, 4	1 k	C1a	470 uF	Q1	MRF961		
R3	100	C2	47 nF	Q2	BC548		
R5	2.2 M	C3a	1 uF	Cv1	2-7 pF miniature		
R6	6.8 k	C4	22 uF	IC1	LM339 comparator		
R7	100 k	C5, 8	100 uF	L1	5 turns 0.86 mm wire on		
R8, 9, 10	22 k	C6	100 nF		3.5 mm core		
Rv1	1 k	C7	10 uF	L2	4 turns 0.86 mm wire on		
					3.5 mm core		

RF DESIGN

Fig. 24-4

The oscillator is a standard UHF design, which delivers about 10 mW at 1.2 GHz. R1 and R2 bias the base of Q1 to 1.2 V via L2. Collector current is set by R3 to about 30 mA. C2 couples the base of Q1 to the stripline circuit. Tuning is provided by CV1, and C1 plus C1a decouple the collector. R2 and R3 are not decoupled because this could cause instability.

Q2 is a simple one-transistor amplifier. C4 and C7 reduce gain below 1.5 and above 100 Hz; the remaining band of frequencies is amplified and passed on to the level detector. Two comparators of IC1 provide level detection. The trigger voltage is set by R7, Rv1, R8, and R9; it is adjustable from 8 to 60 mV by Rv1.

Positive voltage swings above the trigger level cause the IC1a output to become low, while negative swings cause IC1b to become low. C8 decouples IC1 from the power supply, and R10 is a pull-up resistor for the open-collector output of IC1.

CAPACITIVE SENSOR ALARM

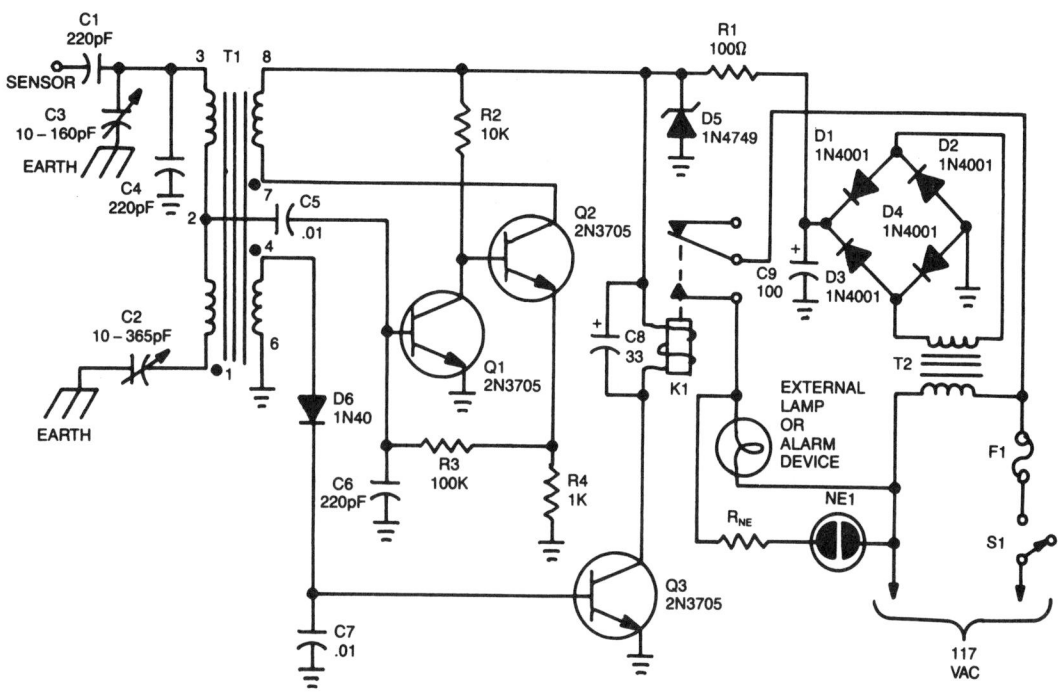

HANDS-ON ELECTRONICS

Fig. 24-5

The unit is constructed around a balanced-bridge circuit, using both capacitance and inductance. The bridge consists of capacitors C2 and C3, and the center-tapped winding of T1. One end of the bridge is coupled to ground by C4, while capacitance changes are introduced through C1. A small capacitance change unbalances the bridge and produces an ac signal at the base of Q1. Transistors Q1 and Q2 are connected to form a modified-Darlington amplifier. The collector load for Q2 is a separate winding of T1 that is connected out-of-phase with the incoming ac signal. That produces a large, distorted signal each time the bridge is unbalanced.

The distorted signal is taken from the bridge circuit by a third winding of transform T1. That signal is then rectified by D6 and applied as a dc signal to the base of Q3. The applied signal energizes the relay, K1, as soon as the unbalanced condition occurs, and the relay drops out as soon as the circuit balance is restored. Of course, for normal alarm use, the relay should be made self-latching so that the alarm condition remains in effect until the system is reset.

An audible alarm, such as a bell or klaxon horn, can be operated from the relay. If a silent alarm is needed, a light bulb can be used. Transformer T1 can be purchased as part #6182 from: Pulse Engineering, P.O. Box 12235, San Diego, CA 92112.

CAPACITANCE RELAY

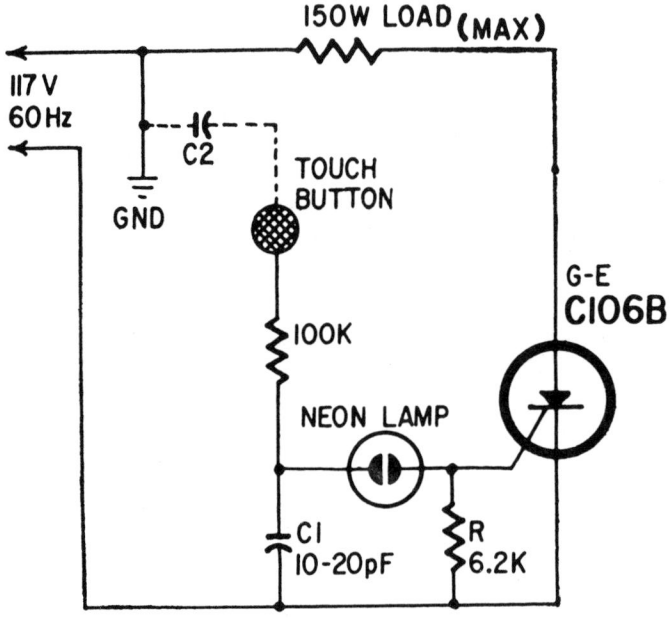

NOTE: ALL RESISTORS 1/2 WATT

RADIO-ELECTRONICS   *Fig. 24-6*

Capacitor C1 and body capacitance (C_2) of the operator form the voltage divider from the hot side of the ac line to ground. The voltage across C1 is determined by the ratio of C_1 to C_2. The higher voltage is developed across the smaller capacitor. When no one is close to the touch button, C_2 is smaller than C_1. When a hand is brought close to the button, C_2 is many times larger than C_1 and the major portion of the line voltage appears across C1. This voltage fires the neon lamp, C1 and C2 discharge through the SCR gate, causing it to trigger and pass current through the load. The sensitivity of the circuit depends on the area of the touch plate. When the area is large enough, the circuit responds to the proximity of an object rather than to touch. C1 can be made variable so that sensitivity can be adjusted.

SELF-BIASED PROXIMITY SENSOR

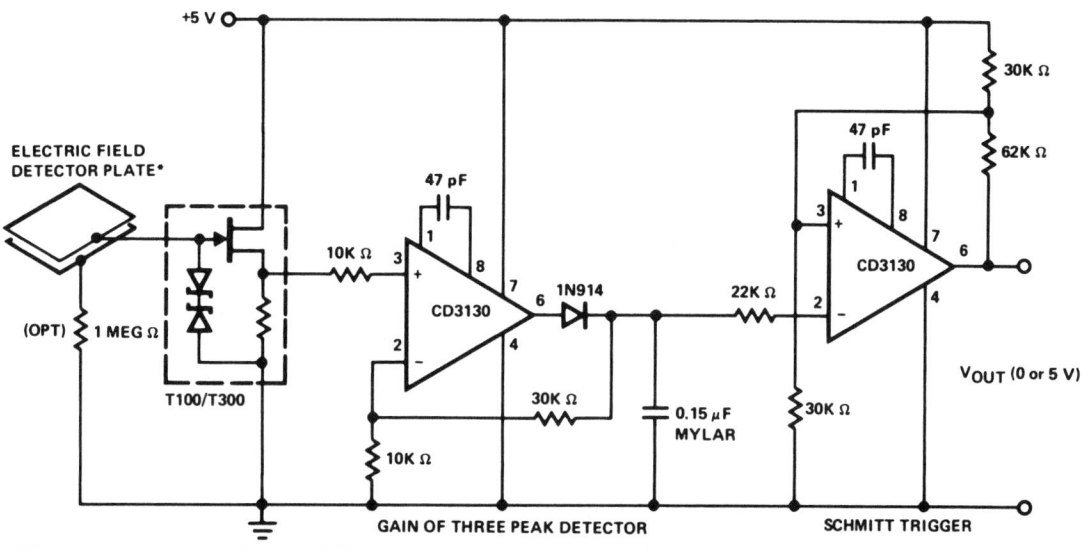

*DETECTOR PLATE MAY BE DOUBLE-SIDED PC BOARD OR ANY INSULATED METAL SHEET

SILICONIX

Fig. 24-7

TOUCH SWITCH OR PROXIMITY DETECTOR

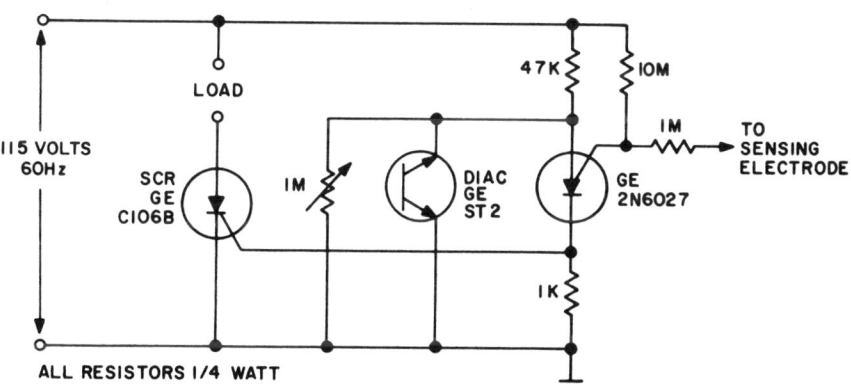

GE

Fig. 24-8

This circuit is actuated by an increase in capacitance between a sensing electrode and the ground side of the line. The sensitivity can be adjusted to switch when a human body is within inches of the insulated plate that is used as the sensing electrode. Thus, sensitivity is adjusted with the 1-MΩ potentiometer which determines the anode voltage level prior to clamping. This sensitivity will be proportional to the area of the surfaces opposing each other.

145

FIELD-DISTURBANCE SENSOR/ALARM

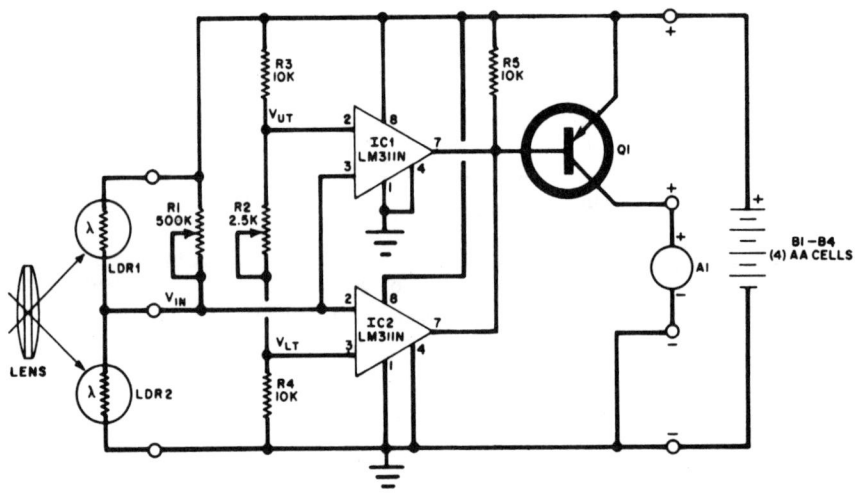

Fig. 24-9

The change in ambient light triggers the alarm by changing resistance of LDR1 and LDR2.

Q1 = Radio Shack 276-2024
A1 = Mallory SC628P Sonalert
LDR1, LDR2 = Cadmium sulfide photocell, Radio Shack 276-116

PROXIMITY DETECTOR

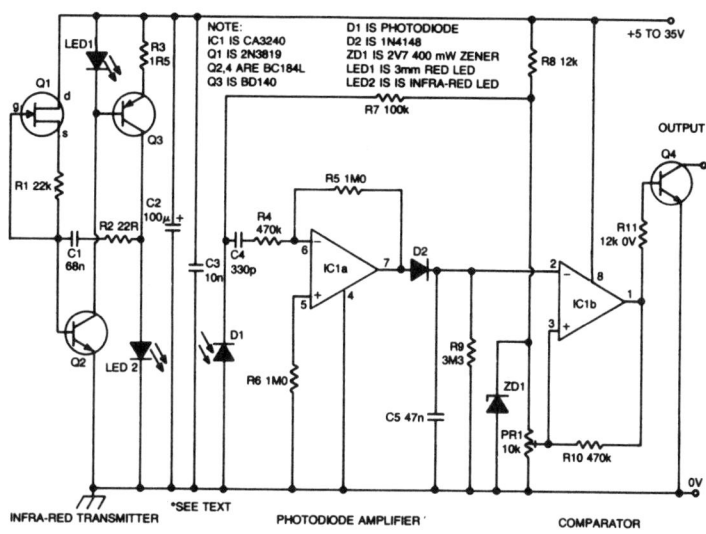

Fig. 24-10

The proximity sensor works on the principle of transmitting a beam of modulated infrared light from the emitter diode LED2, and receiving reflections from objects passing in front of the beam with a photodiode detector D1. The circuit can be split into three distinct stages; the infrared transmitter, the photodiode amplifier, and a variable-threshold comparator.

PROXIMITY SENSOR

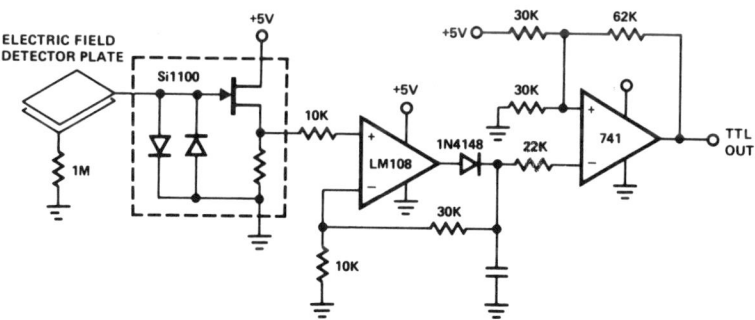

Fig. 24-11

The Si1100 series circuit input is connected to a capacitive field sensor—possibly a piece of double-sided circuit board. Any induced voltage change on the plate is fed to the input of the peak detector section of the op-amp circuit. The Schmitt trigger monitors the voltage across the capacitor and changes its output state when the capacitor voltage crosses the 2.5 trigger point. The output from the Schmitt trigger switches between 0 and 5 V and is microprocessor compatible for sensor applications, such as computer-controlled intruder alarms.

25

Pulse Detectors

The sources of the following circuits are contained in the Sources section, which begins on page 214. The figure number in the box of each circuit correlates to the source entry in the Sources section.

Digital Frequency Detector   Missing-Pulse Detector I
Pulse-Sequence Detector   Missing-Pulse Detector II
Pulse-Width Discriminator   Out-of-Bounds Pulse-Width Detector
Edge Detector   Pulse-Coincidence Detector

DIGITAL FREQUENCY DETECTOR

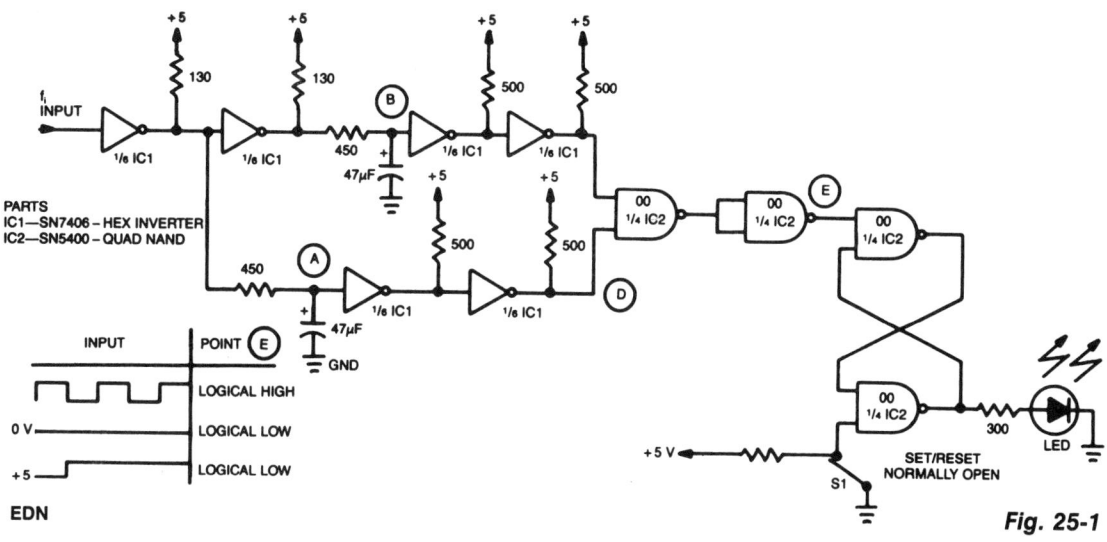

EDN

Fig. 25-1

A simple inverter and a NAND gate can be connected to yield a highly compact and reliable digital frequency detector. This circuit can detect frequencies up to 3 MHz with 50% duty cycles. When a frequency, f_i, appears at the input, points A and B detect a logical high dc level. Thereupon point E increases the latch sets and the LED lights. If the input frequency is absent and if the voltage is either at a constant high or low level, points A and B will be complementary and point E will decrease. This will reset the latch and extinguish the LED.

PULSE-SEQUENCE DETECTOR

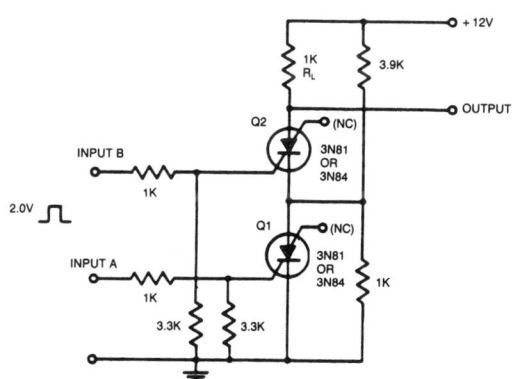

The resistor divider connected between Q1 and Q2 supplies I_H to Q1 after input A triggers it. It also prevents input B from triggering Q2 until Q1 conducts. Consequently, the first B input pulse after input A is applied, will supply current to R_L.

GE

Fig. 25-2

PULSE-WIDTH DISCRIMINATOR

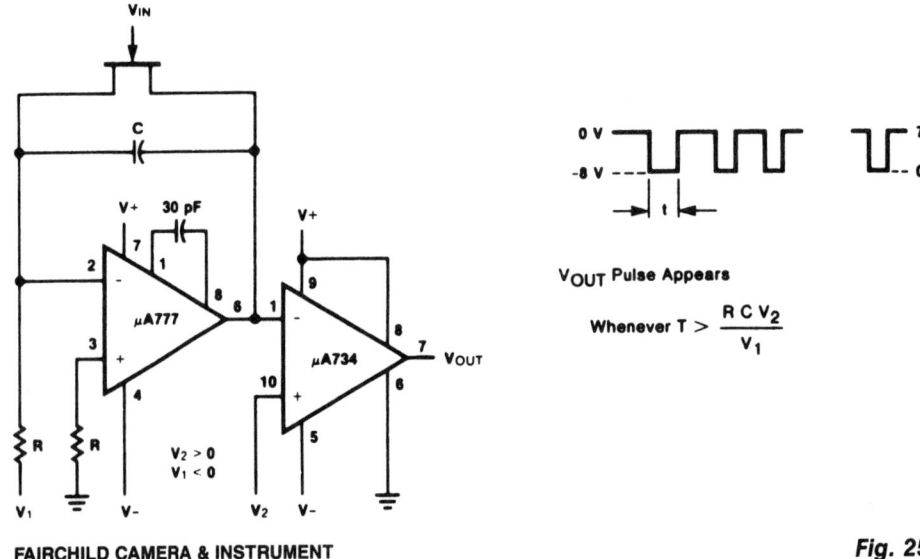

V_{OUT} Pulse Appears

Whenever $T > \dfrac{R C V_2}{V_1}$

FAIRCHILD CAMERA & INSTRUMENT

Fig. 25-3

EDGE DETECTOR

RETRIGGERABLE
MONOSTABLE
a

$T = 0.8RC$

HALF
MONOSTABLES
b

RADIO-ELECTRONICS   *Fig. 25-4*

The 555 is a monostable that *wants* a negative-going trigger. If the pulse you're feeding it with is positive-going, you can run it through an inverter made up of either an inverting gate or, if you're tight on space, a single transistor. Both ways are shown. The circuits shown in Fig. 25-4b are edge detectors as well, and are usually referred to as *half monostables*, because they can't be used in every application. The width of the output pulse is determined by the RC value, but a few rules govern their use:

- The input pulse must be wider than the output pulse
- The input pulse can't be glitchy
- The circuit can't be retriggered faster than the RC time

MISSING-PULSE DETECTOR I

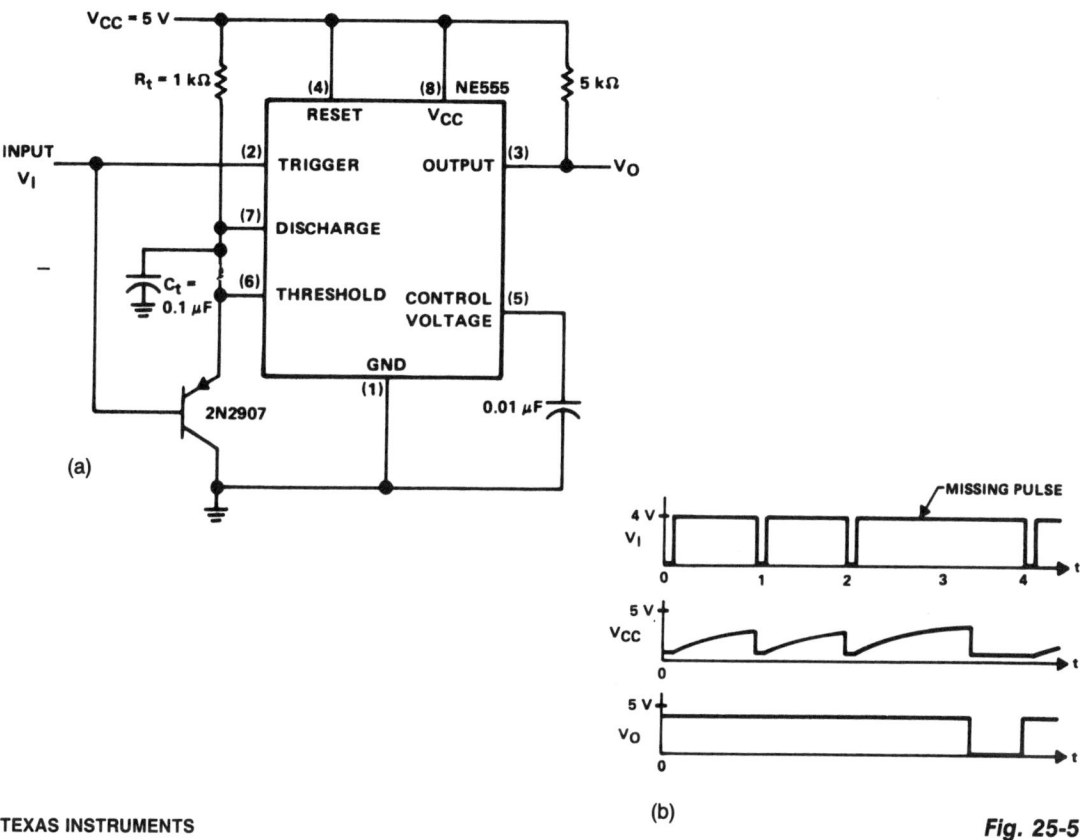

(a)

(b)

Fig. 25-5

This circuit will detect a missing pulse or an abnormally long spacing between consecutive pulses in a train of pulses. The timer is connected in the monostable mode. The time delay should be set slightly longer than the timing of the input pulses. The timing interval of the monostable circuit is continuously retriggered by the input pulse train, V_I. The pulse spacing is less than the timing interval, which prevents V_C from rising high enough to end the timing cycle. A longer pulse spacing, a missing pulse, or a terminated pulse train will permit the timing interval to be completed. This will generate an output pulse, V_O as illustrated in Fig. 25-5b. The output remains high on pin 3 until a missing pulse is detected, at which time the output decreases.

The NE555 monostable circuit should be running slightly slower and lower in frequency than the frequency to be analyzed. Also, the input cannot be more than twice this free-running frequency or it would retrigger before the timeout and the output would remain in the low state continuously. The circuit operates in the monostable mode at about 8 kHz, so pulse trains of 8 to 16 kHz can be observed.

MISSING-PULSE DETECTOR II

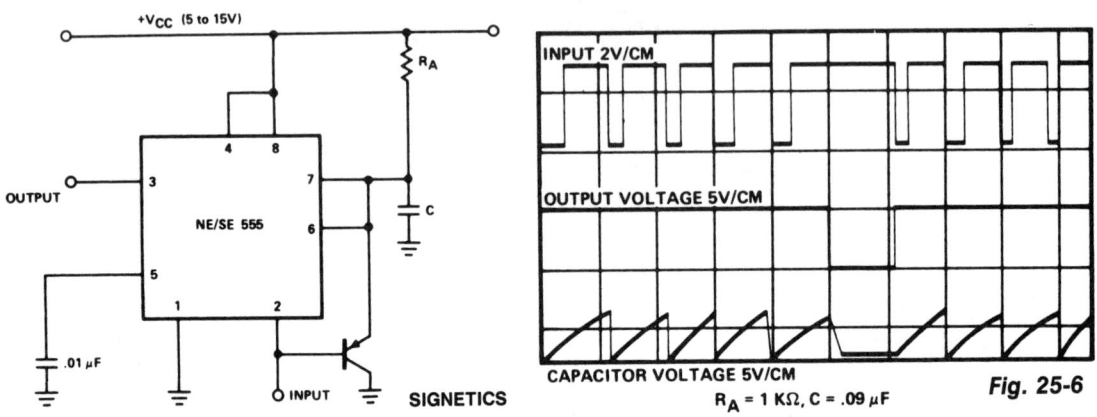

Fig. 25-6

The timing cycle is continuously reset by the input pulse train. A change in frequency or a missing pulse allows completion of the timing cycle, which causes a change in the output level. For this application, the time delay should be set to be slightly longer than the normal time between pulses. The graph shows the actual waveforms seen in this mode of operation.

OUT-OF-BOUNDS PULSE-WIDTH DETECTOR

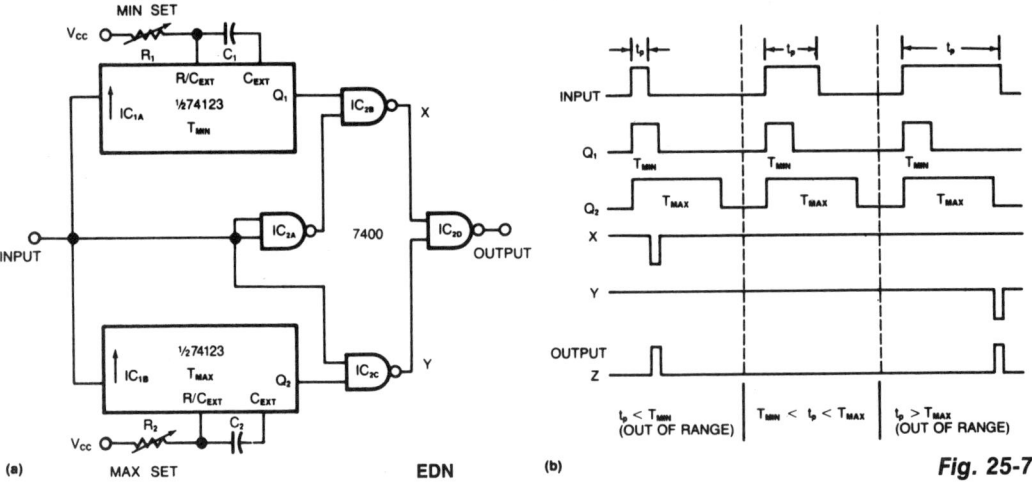

Fig. 25-7

Requiring only two ICs to monitor a train of positive pulses, this circuit produces a single positive output pulse for each input pulse, whose duration is either too long or too short. You specify the minimum and maximum limits by adjusting the trimming potentiometers, R1 and R2. You can set the value of the acceptable pulse width from approximately 50 ns to 10 μs for a 74123 monostable multivibrator. The leading edge of an input pulse triggers one shots IC1A and IC1B, as you can see from the timing diagram. Each NAND-gate output is high, unless either or both inputs are low, so outputs X and Y are high unless the circuit encounters an out-of-range pulse. IC2D then gates a negative pulse from IC2B or IC2C to produce the circuit's positive output pulse.

PULSE-COINCIDENCE DETECTOR

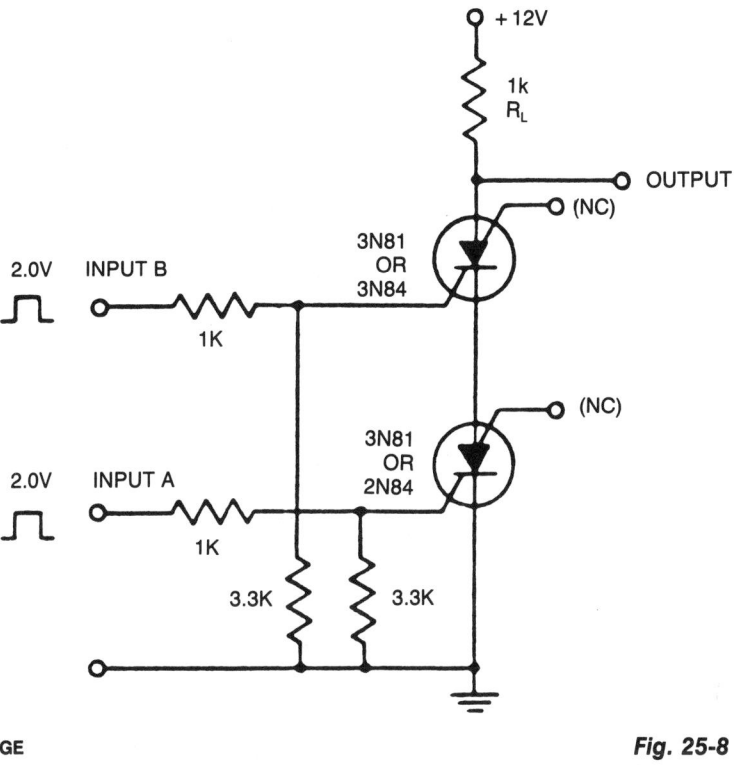

Fig. 25-8

Unless inputs A and B (2- to 3-V amplitude) occur simultaneously, no voltage exists across R_L. Less than 1-μs overlap is sufficient to trigger the SCS. Coincidence of negative inputs is detected with gate G_A instead of G_C by using the SCS in a complementary SCR configuration.

26

Radar Detectors

The sources of the following circuits are contained in the Sources section, which begins on page 214. The figure number in the box of each circuit correlates to the source entry in the Sources section.

ONE-CHIP RADAR DETECTOR

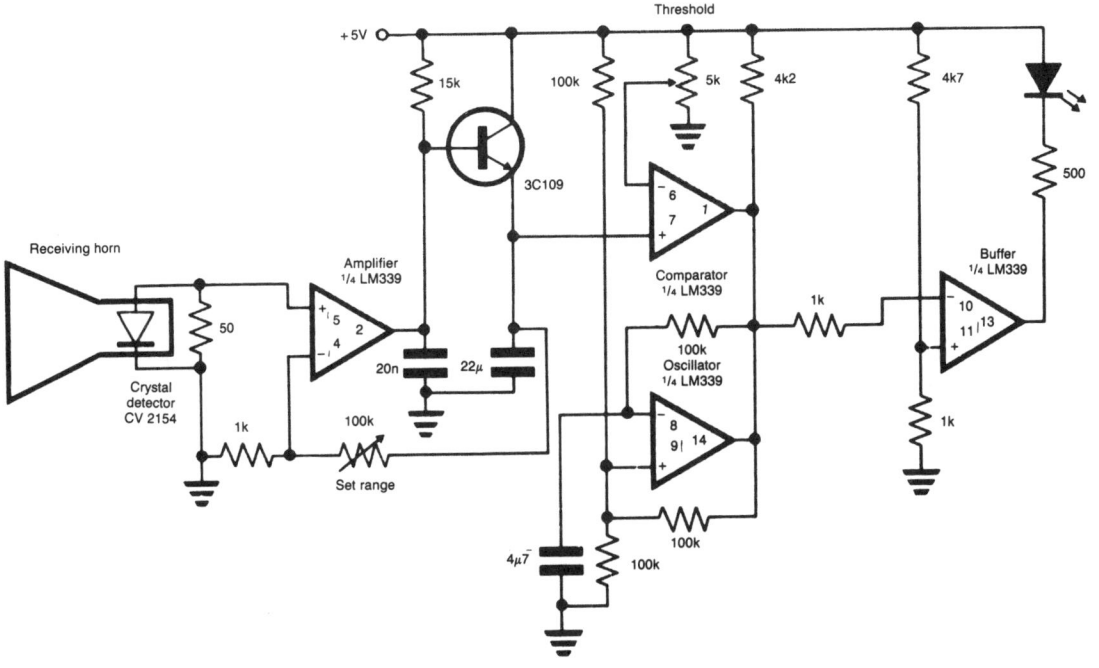

Fig. 26-1

A simple X-band radar detector is capable of indicating changes in RF radiation strength at levels down to 2 mW/cm^2. Radiation falling on the detector diode, produces a voltage at the input of an amplifier whose gain can be adjusted to vary the range at which the warning is given. The amplifier output drives a voltage comparator with a variable threshold set to a level that avoids false alarms. The comparator output is connected in the wired-OR configuration, with the open-collector output of an oscillator running at a frequency of 2 Hz. In the absence of a signal, the comparator output level is low, which inhibits the oscillator output stage and holds the buffer so that the lamp is off. When a signal appears, the comparator output goes high, removes the lock from the oscillator (which free-runs), and switches the lamp on and off at 2 Hz.

RADAR-SIGNAL DETECTOR

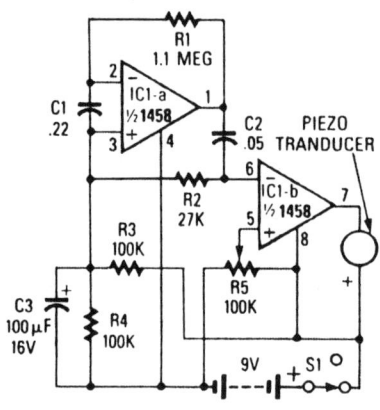

FIG. 1—THE ECONOMY RADAR DETECTOR needs only one IC and a few discrete components.

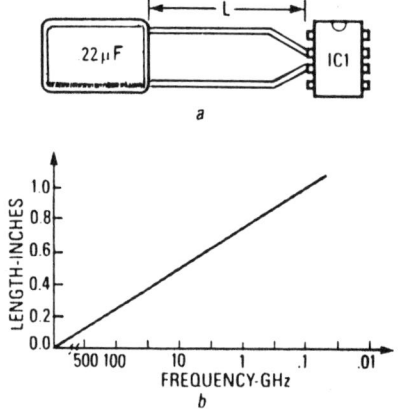

FIG. 3—VARY THE LEAD LENGTHS OF C1 to tune the input circuit.

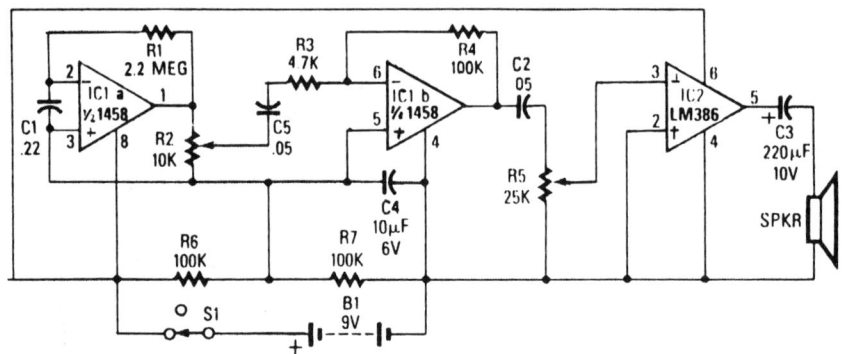

FIG. 2—DELUXE RADAR DETECTOR adds a buffer amplifier and an audio power amp to drive a speaker.

RADIO-ELECTRONICS

Fig. 26-2

The circuit can be tuned to respond to signals between 50 MHz and 500 GHz. The economy model is shown in Fig. 1, and the deluxe model is shown in Fig. 2. The first op amp in each circuit functions as a current-to-voltage converter. In the economy model, IC1b buffers the output to drive the piezo buzzer. The deluxe model functions in a similar manner, except that IC1b is configured as a ×20 buffer amplifier to drive the LM386. In both circuits, C1 functions as a "transmission line" that intercepts the incident radar signal. The response can be optimized by trimming C1's lead length for the desired frequency. Typically, the capacitor's leads should 0.5 to 0.6″ long.

27

Radiation Detectors

The sources of the following circuits are contained in the Sources section, which begins on page 214. The figure number in the box of each circuit correlates to the source entry in the Sources section.

MICROPOWER RADIOACTIVE RADIATION DETECTOR

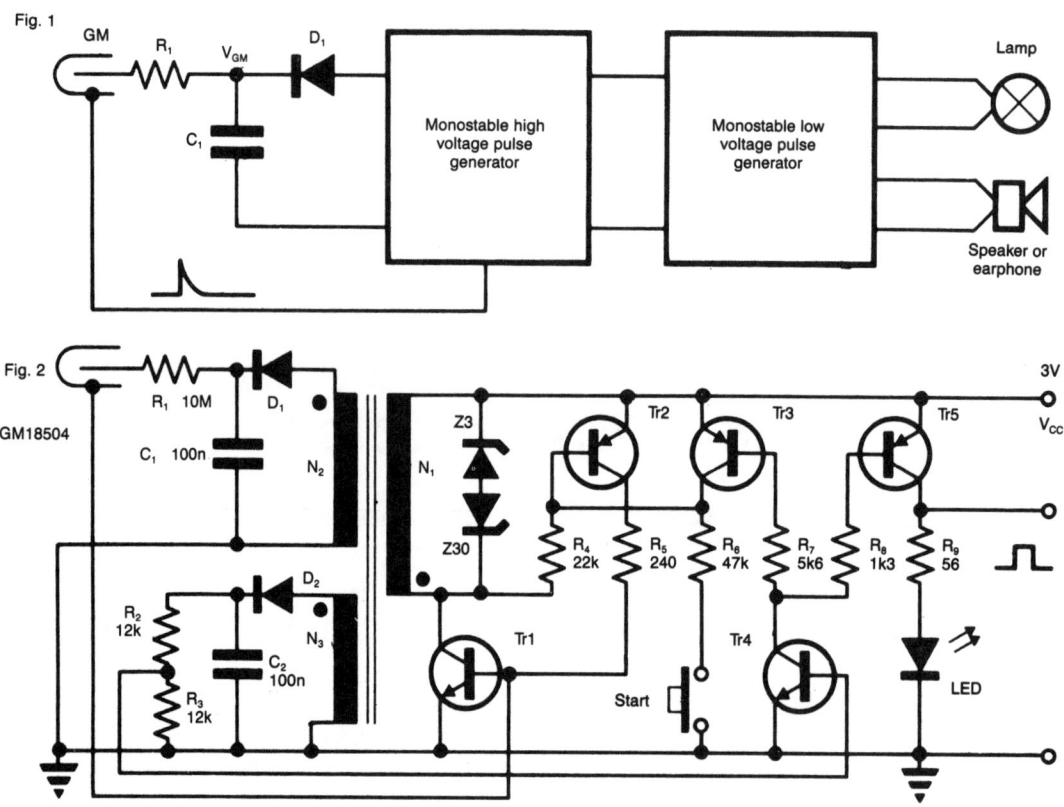

Fig. 27-1

In the absence of radiation, no current is drawn. At normal background radiation levels the power consumption is extremely low. The instrument can be left on for several months without changing batteries. In this way, the detector is always ready to indicate an increase in radiation. An LED is used as an indicator lamp. With background radiation, it draws less than 50 μA. A ferrite pot core is used for the transformer with $N_1 = 30$, $N_2 = 550$, and $N_3 = 7$. Using two 1.5-V batteries with 0.5-Ah total capacity, the detector can work at background radiation levels for 0.5 Ah ÷ 50 μA = 10,000 hours, which is more than a year.

WIDEBAND RADIATION MONITOR

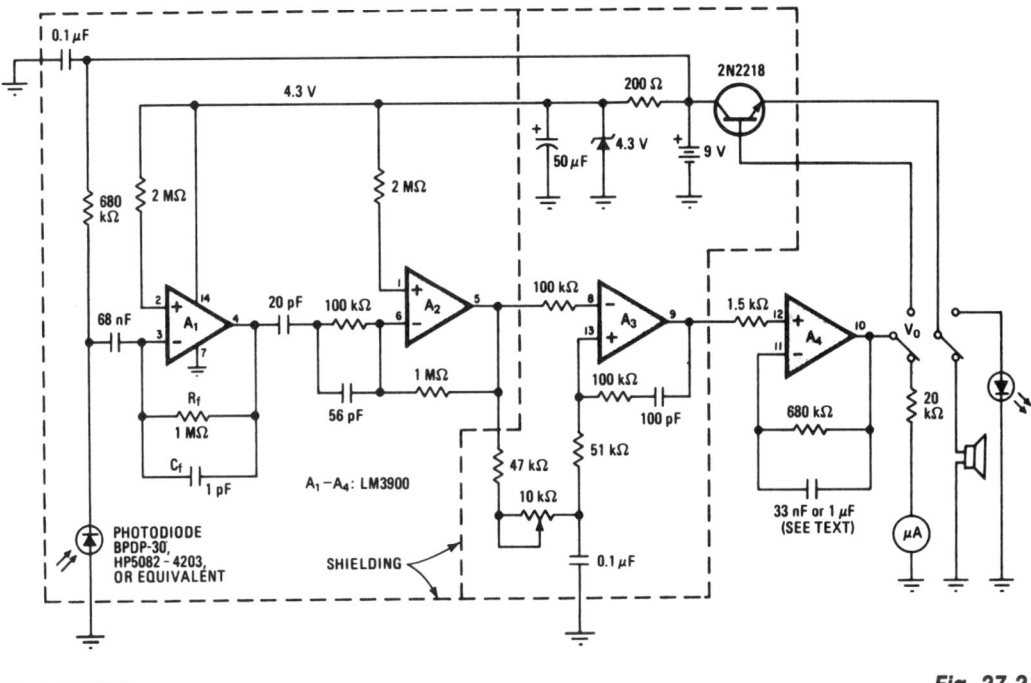

Fig. 27-2

A sensitive radiation monitor can be simply constructed with a large-area photodiode and a quad operational amplifier. Replacing the glass window of the diode with Mylar foil will shield it from light and infrared energy, enabling it to respond to such nuclear radiation as alpha and beta particles and gamma rays. A4 integrates the output of A3 in order to drive a microammeter. A 1-μF capacitor is used in the integrating network. A lower value, say, 33 nF, will make it possible to drive a small loudspeaker (50-Hz output signal) or LED.

POCKET-SIZED GEIGER COUNTER

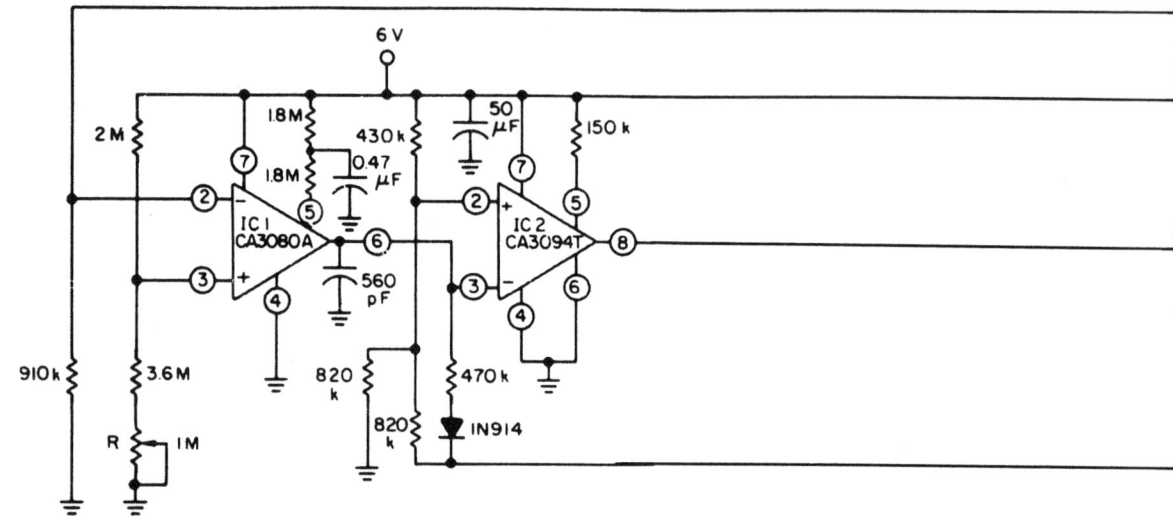

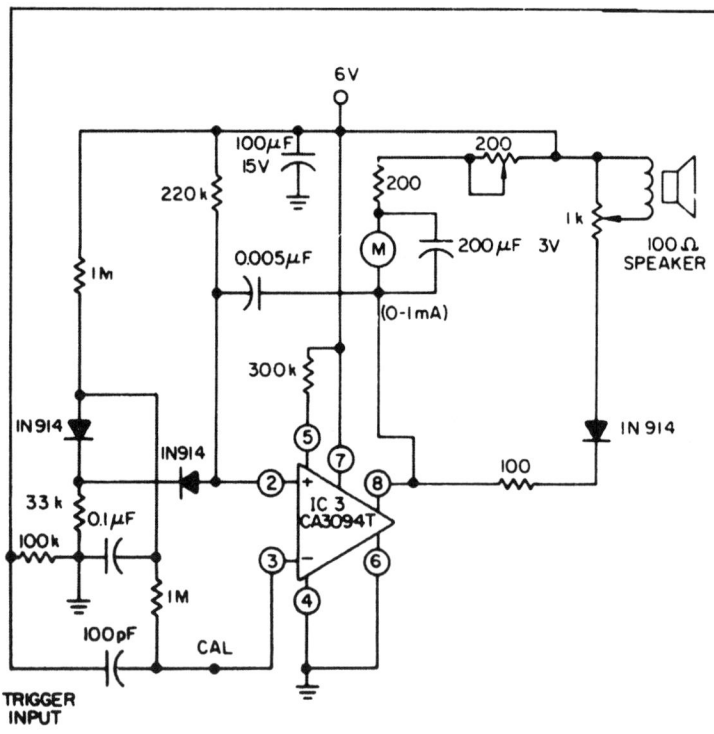

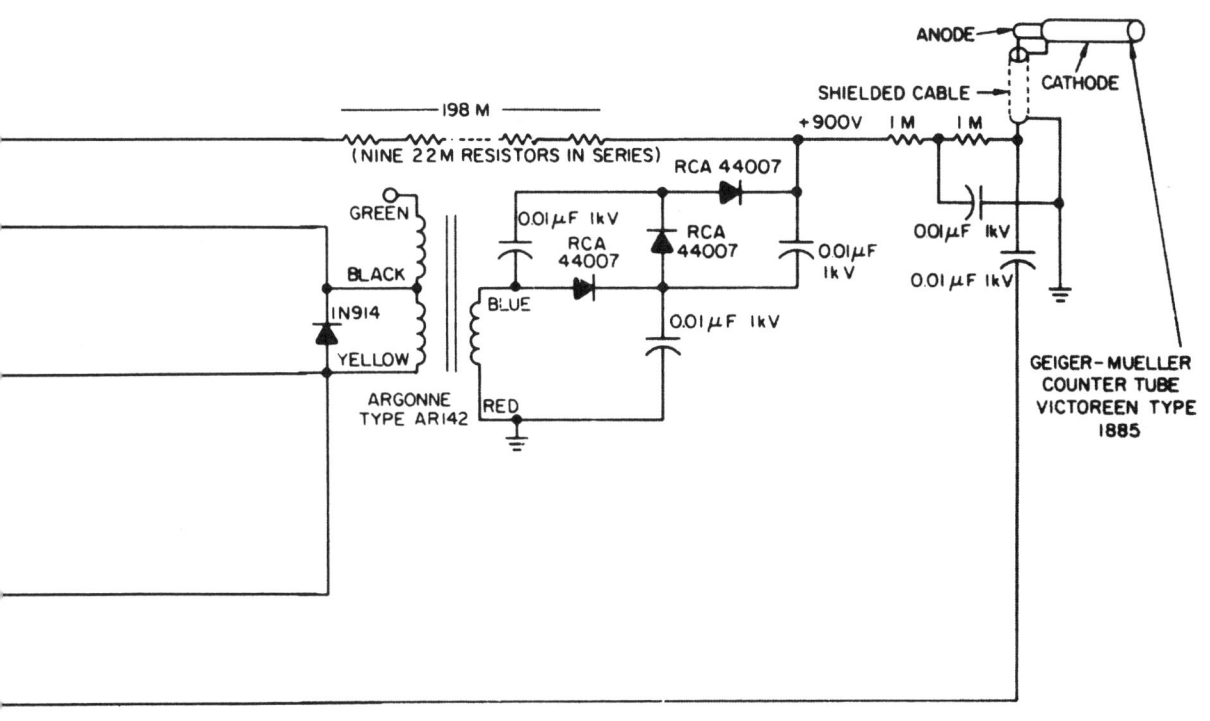

A single 6.75-V mercury battery powers the counter, which features a 1-mA count-rate meter, as well as an aural output. A regulated 900-V supply provides stable operation for the counter tube. A multivibrator, built around a differential power amplifier IC2, drives the step-up transformer. Comparator IC1 varies the multivibrator duty cycle to provide a constant 900 V. The entire regulated supply draws less than 2 mA. A one-shot multivibrator, built with IC3, provides output pulses that have constant width and amplitude. Thus, the average current through the meter is directly proportional to the pulse-rate output from the counter tube. And the constant-width pulses also drive the speaker. Full-scale meter deflection (1 mA) represents 5000 counts/min, or 83.3 pulses/s. A convenient calibration checkpoint can be provided on the meter scale for 3600 ppm (60 pulses/s).

PHOTOMULTIPLIER OUTPUT-GATING CIRCUIT

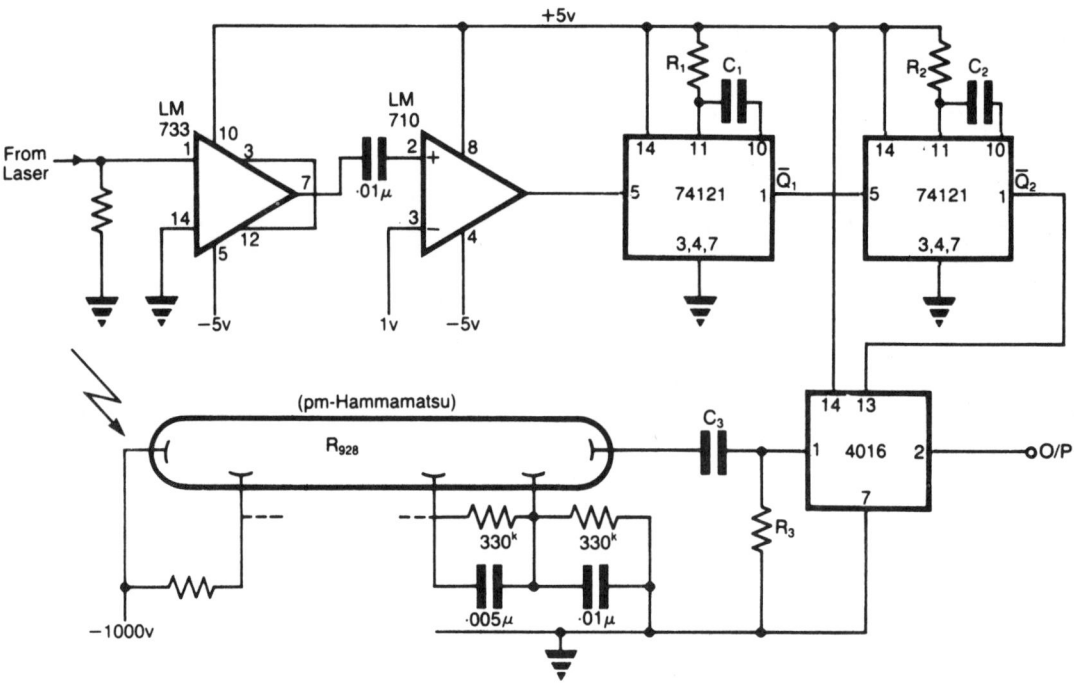

Fig. 27-4

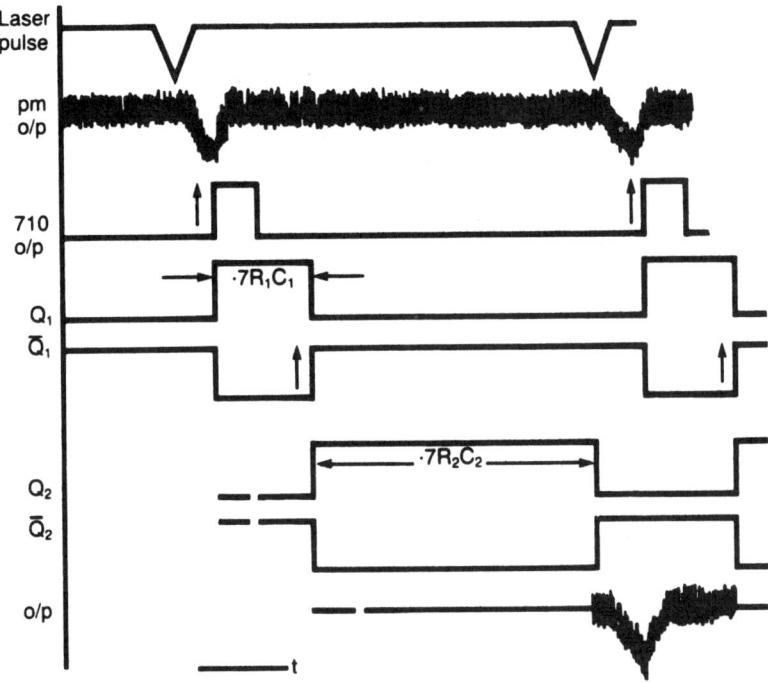

The application involves observing the light pulse emerging from a thick specimen after transillumination by a laser pulse. Pulses derived from the laser source are amplified using a video amplifier LM733. The reference level is set to 1 V in the comparator LM710, to provide the necessary trigger pulses for the monostable multivibrator 74121. The laser pulses have a repetition frequency of 500 Hz and suitable values are:

$R_1 = 33$ kΩ, $C_1 = 22$ pF
$R_2 = 33$ kΩ, $C_2 = 68$ nF

The pulse width for each monostable is approximately given by $pw = 0.7\,RC$. R3/C3 is a high-pass filter. This method therefore permits the use of low-cost components that have moderate response times for extracting the pulse of interest.

GAMMA-RAY PULSE INTEGRATOR

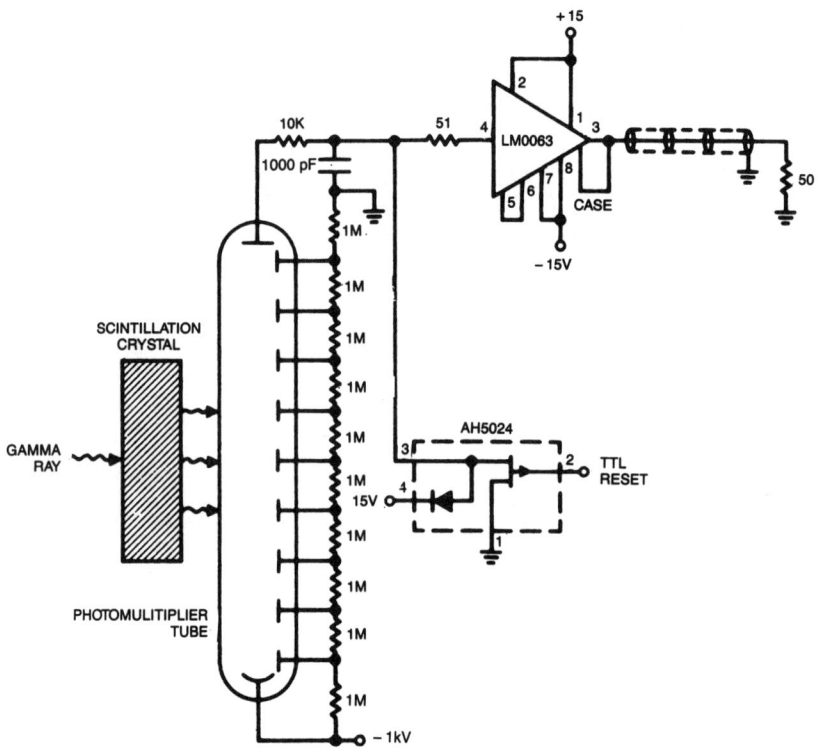

NATIONAL SEMICONDUCTOR

Fig. 27-5

SENSITIVE GEIGER COUNTER

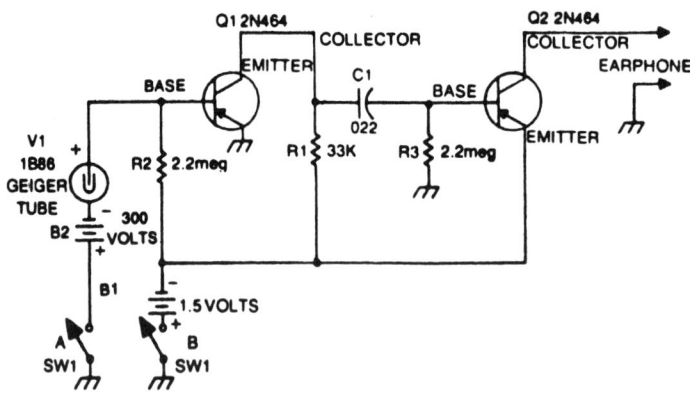

TAB BOOKS

Fig. 27-6

GEIGER COUNTER

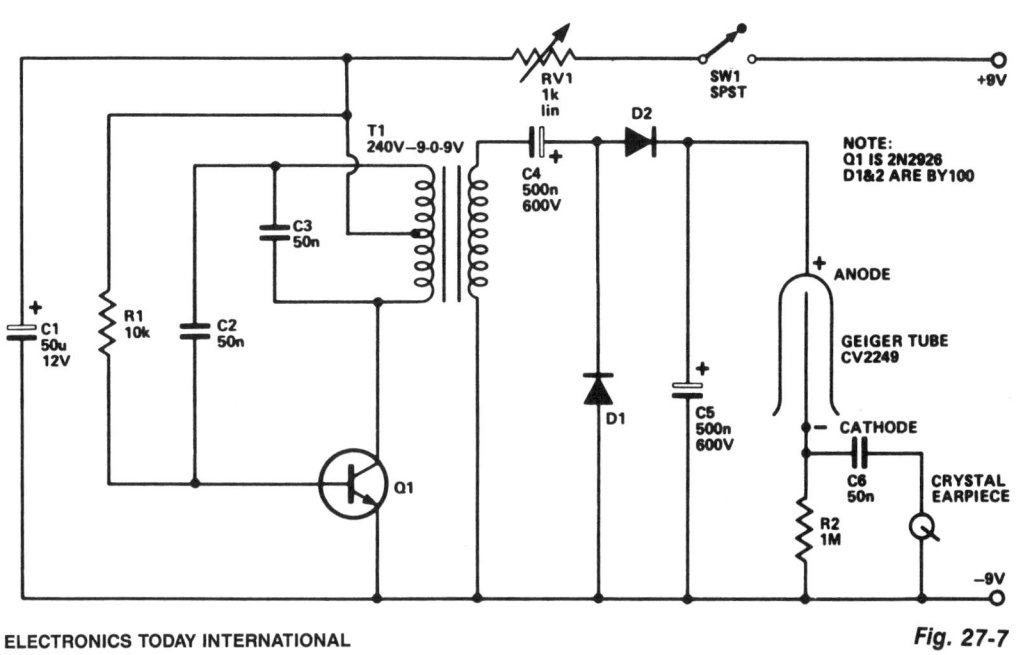

ELECTRONICS TODAY INTERNATIONAL

Fig. 27-7

The Geiger tube needs a high-voltage supply, which consists of Q1 and its associated components. The transformer is connected in reverse; the secondary is connected as a Hartley oscillator, and R1 provides base bias. D1, D2, C4, and C5 compose a voltage doubler. RV1 should be set so that each click heard is nice and clean—over a certain voltage range all that will be heard is a continuous buzz.

NUCLEAR PARTICLE DETECTOR

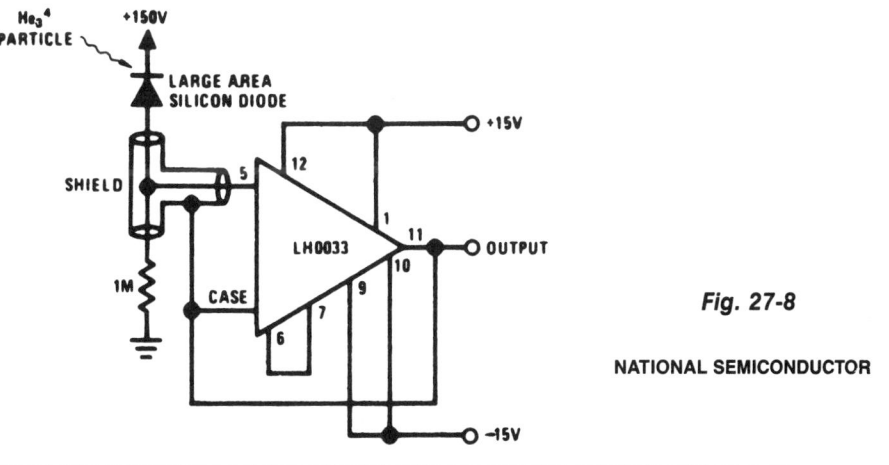

Fig. 27-8

NATIONAL SEMICONDUCTOR

28

Rain Detectors

The sources of the following circuits are contained in the Sources section, which begins on page 214. The figure number in the box of each circuit correlates to the source entry in the Sources section.

Rain-Warning Bleeper
Rain Alarm/Doorbell
Acid Rain Monitor
Rain Alarm

RAIN-WARNING BLEEPER

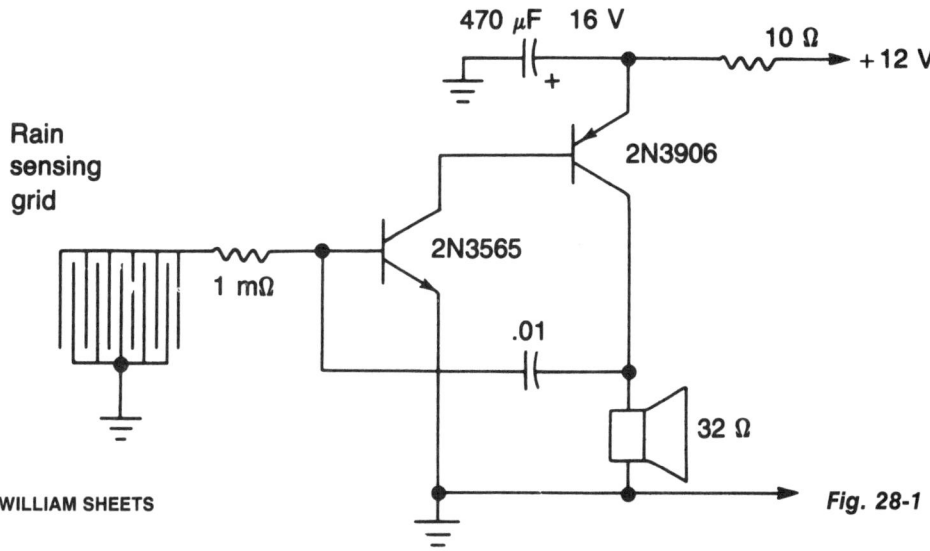

WILLIAM SHEETS

Fig. 28-1

One small spot of rain on the sense pad of this bleeper will start this audio warning. It can also be operated by rising water. The circuit has two transistors, with feedback via capacitor C1, but Tr1 cannot operate as long as the moisture sense pad is dry. When the pad conducts (Tr1 and Tr2) form an audio oscillator circuit. The pitch depends somewhat on the resistance value.

RAIN ALARM/DOORBELL

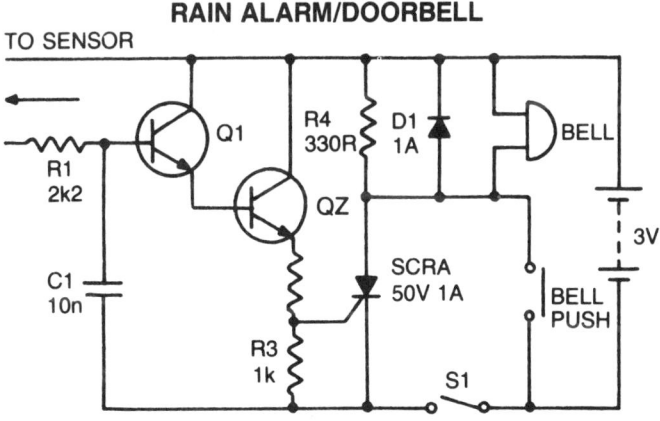

ELECTRONICS TODAY INTERNATIONAL

Fig. 28-2

With S1 open, the circuit functions as a doorbell. With S1 closed, rain falling on the sensor will turn on Q1, trigger Q2 and the thyristor, and activate the bell; R4 provides the holding for the thyristor and D1 prevents any damage to the thyristor from the back of EMF in the bell coil. The sensor can be made from 3 square inches of copper-clad board with a razor cut down the center. C1 prevents any main pickup in the sensor leads.

ACID RAIN MONITOR

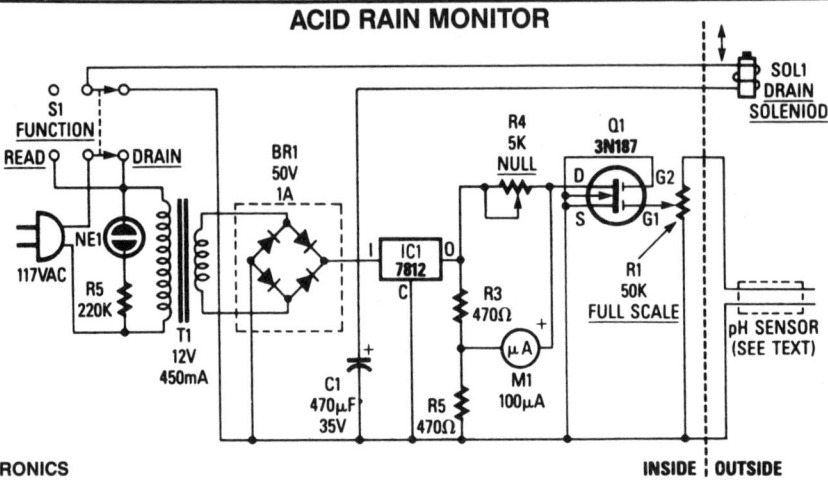

RADIO-ELECTRONICS

INSIDE | OUTSIDE

Fig. 28-3

A bridge rectifier and 12-V regulator powers the MOSFET sensing circuit. The unregulated output of the bridge rectifier operates the drain solenoid via switch S1. The sensor itself is built from two electrodes, one made of copper, the other of lead. In combination with the liquid trapped by the sensor, they form a miniature lead-acid cell, whose output is amplified by MOSFET Q1. The maximum output produced by our prototype cell was about 50 μA. MOSFET Q1 serves as the fourth leg of a Wheatstone bridge. When acidity causes the sensor to generate a voltage, Q1 turns on slightly, so its drain-to-source resistance decreases. That resistance variation causes an imbalance in the bridge, and that imbalance is indicated by meter M1.

RAIN ALARM

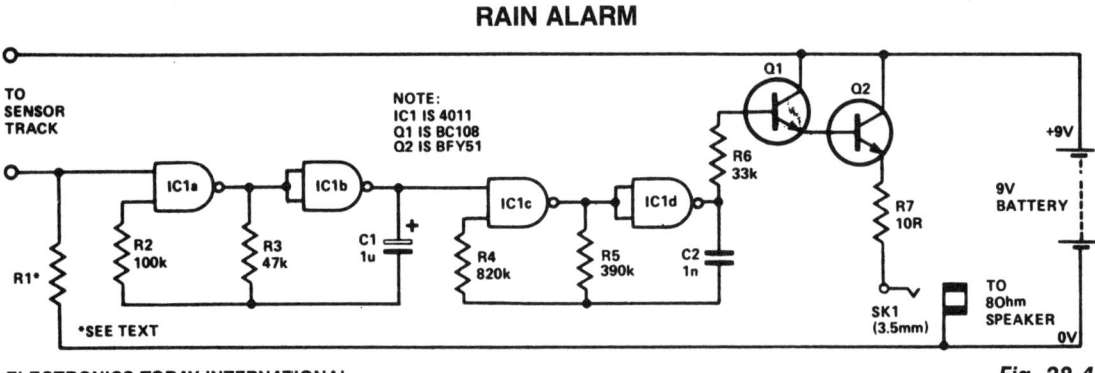

ELECTRONICS TODAY INTERNATIONAL

Fig. 28-4

The circuit uses four NAND gates of a 4011 package. In each oscillator, while one gate is configured as a straightforward inverter, the other has one input that can act as a control input. Oscillator action is inhibited if this input is held low. The first oscillator (IC1a and IC1b) has this input tied low via a high-value resistor (R1) that acts as a sensitivity control. Thus, this oscillator will be disabled until the control input is taken high. Any moisture bridging the sensor track will so enable the output, which is a square wave at about 10 Hz. This, in turn, will gate on and off the 500-Hz oscillator formed by IC1c and IC1d. This latter oscillator drives the loudspeaker via R6, the Darlington pair formed by Q1 and Q2, and resistor R7.

29

Stereo Balance Detectors

The sources of the following circuits are contained in the Sources section, which begins on page 214. The figure number in the box of each circuit correlates to the source entry in the Sources section.

Stereo Balance Tester
Stereo Balance Meter

STEREO BALANCE TESTER

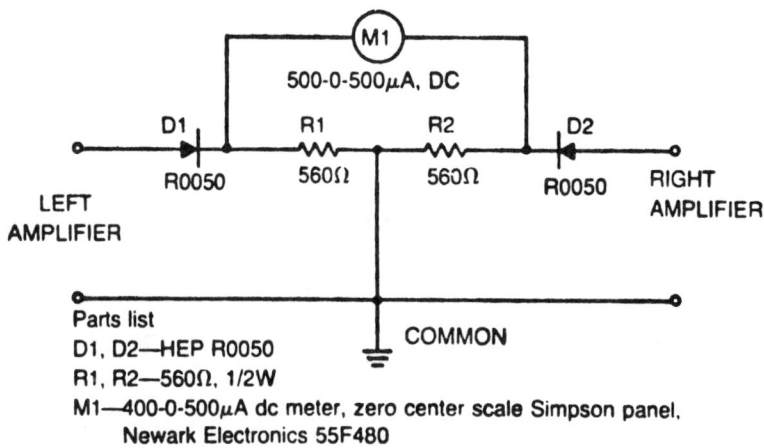

Parts list
D1, D2—HEP R0050
R1, R2—560Ω, 1/2W
M1—400-0-500μA dc meter, zero center scale Simpson panel,
   Newark Electronics 55F480

TAB BOOKS

Fig. 29-1

The meter will show volume- and tone-control balance between left and right stereo amplifiers. For maximum convenience, the meter is a zero-center type. Resistors are five percent or better and the diodes a matched pair. Optimum stereo level and phase balance occurs for matched speakers when the meter indicates zero. If the meter indicates either side of zero, the levels are not matched or the wires are incorrectly phased. Check phasing by making certain that the meter leads are connected to the amplifier hot terminals and that the common leads go to ground.

STEREO BALANCE METER

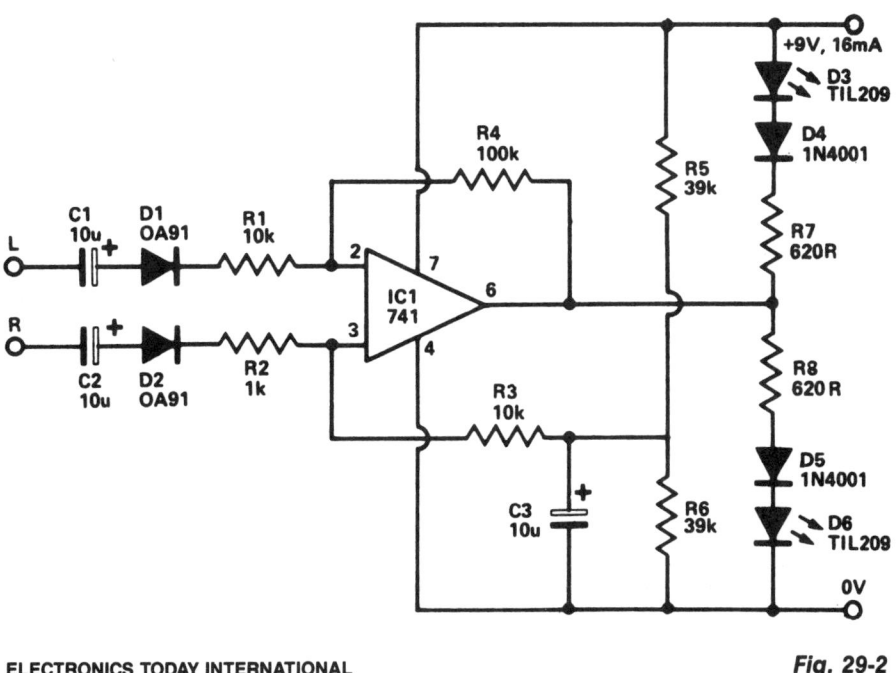

ELECTRONICS TODAY INTERNATIONAL

Fig. 29-2

To use the indicator, switch the amplifier to the mono mode and adjust the balance control until both LEDs are equally illuminated. The amplifier is now in perfect stereo-mode balance.

30

Telephone Detectors

The sources of the following circuits are contained in the Sources section, which begins on page 214. The figure number in the box of each circuit correlates to the source entry in the Sources section.

Tone Dial-Sequence Decoder
Remote Ring-Extender Switch
Tone Dial Decoder
Telephone Relay
Speech-Activity Detector for Telephone Lines

Ring Detector
Telephone Off-Hook Indicator
Frequency- and Volume-Controlled Telephone
    Ring Detector

TONE DIAL-SEQUENCE DECODER

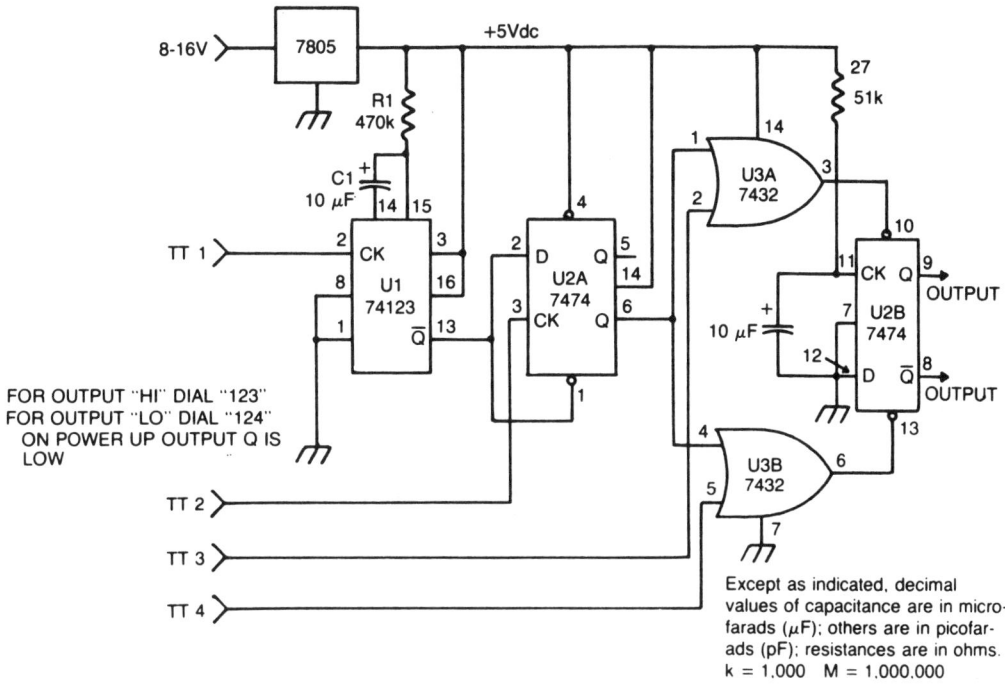

HAM RADIO

Fig. 30-1

The circuit takes active low inputs from a touch tone decoder and reacts to a proper sequence of digits. The proper sequence is determined by which touch tone digits the user connects to the sequence-decoder inputs TT1, TT2, TT3, and TT4.

REMOTE RING-EXTENDER SWITCH

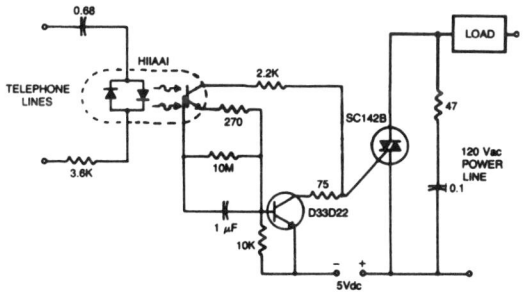

The circuit can operate lamps and buzzers from the 120-V 60-Hz power line, while maintaining positive isolation between the telephone line and the power line. Use of the isolated tab triac simplifies heatsinking by removing the constraint of isolating the triac heatsink from the chassis.

GE

Fig. 30-2

TONE DIAL DECODER

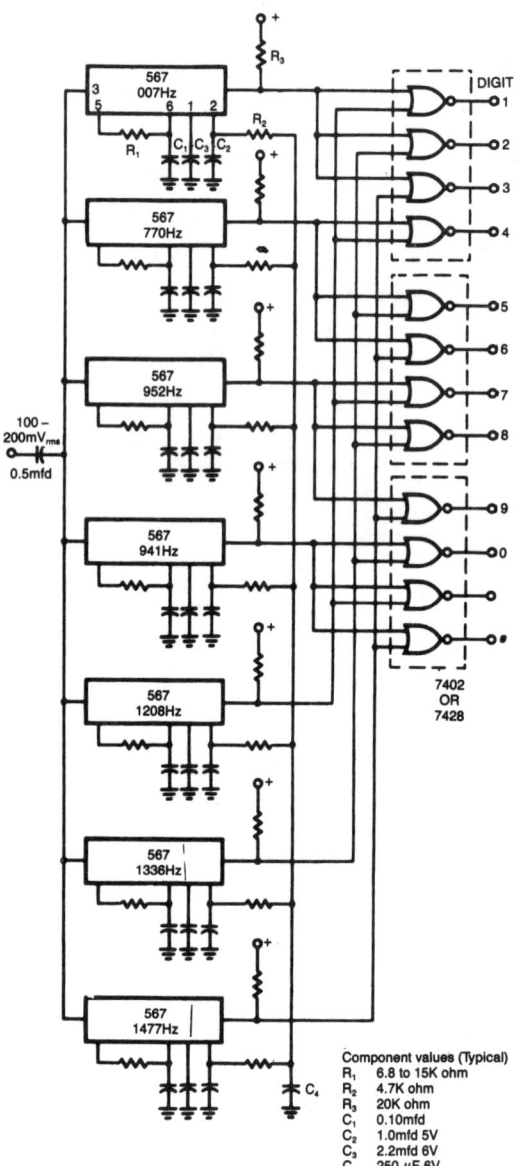

100 –
200mV$_{rms}$
0.5mfd

Component values (Typical)
R$_1$ 6.8 to 15K ohm
R$_2$ 4.7K ohm
R$_3$ 20K ohm
C$_1$ 0.10mfd
C$_2$ 1.0mfd 5V
C$_3$ 2.2mfd 6V
C$_4$ 250 μF 6V

SIGNETICS

Fig. 30-3

TELEPHONE RELAY

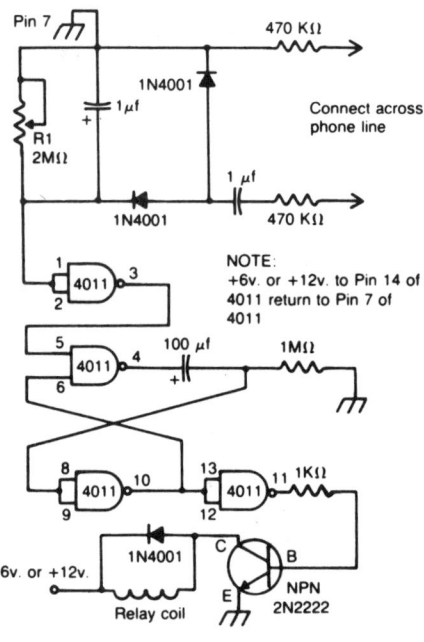

MODERN ELECTRONICS   *Fig. 30-4*

Connected across the bell circuit of phone, this circuit closes a relay when the phone is ringing. Use the delay contacts to actuate any bell, siren, buzzer, or lamp.

SPEECH-ACTIVITY DETECTOR FOR TELEPHONE LINES

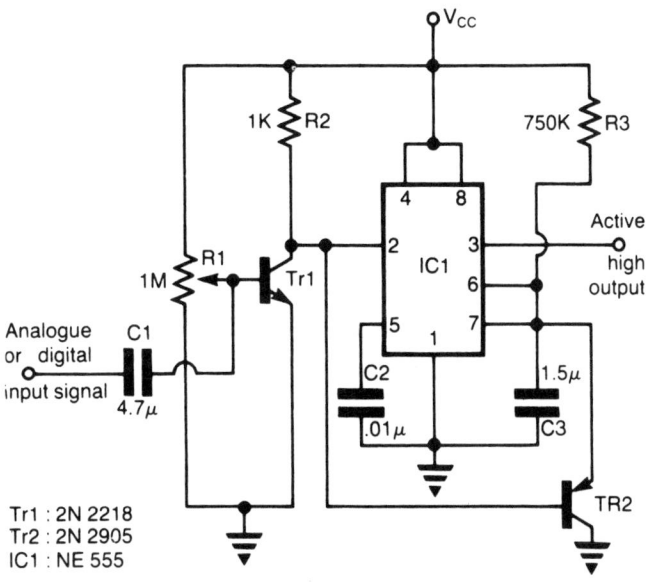

ELECTRONIC ENGINEERING   *Fig. 30-5*

The circuit can be used in telephone lines for speech-activity detection purposes. This detection is very useful in the case of half-duplex conversation between two stations (in the case of simultaneous transmission of voice and data over the same pair of cables by the method of interspersion data on voice traffic) and also in echo-suppressor devices. The circuit consists of a class-A amplifier in order to amplify the weak analog signals (in the range of 25 to 400 mW of an analog telephone line).

The IC1 is connected as a retriggerable monostable multivibrator with Tr2 discharging the timing capacitor (C3), if the pulse train reaches the trigger input 2 of IC1 with period less than the time:

$$T_{high} = 1.1(R_3\,C_3)$$

The output 3 of IC1 is active ON when an analog or digital signal is presented at the output and it drops to low level, T_{high} seconds after the input signal has ceased to exist.

RING DETECTOR

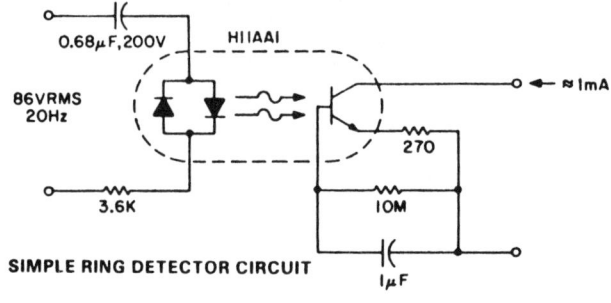

SIMPLE RING DETECTOR CIRCUIT

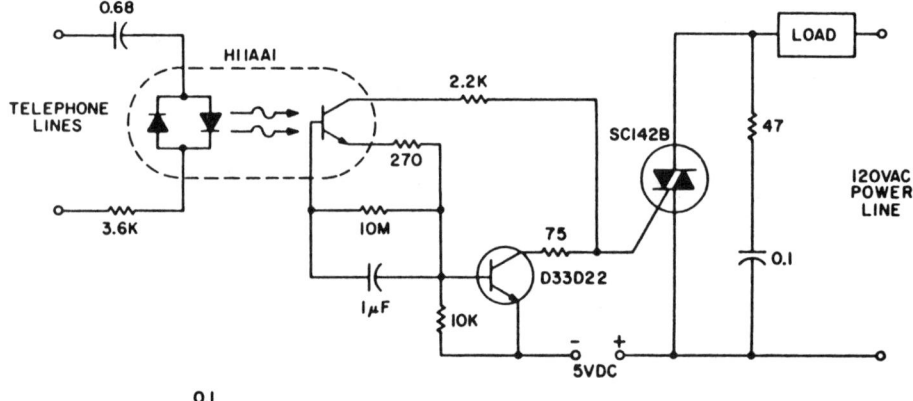

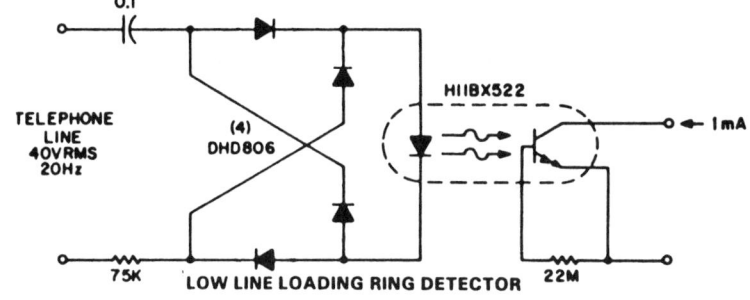

LOW LINE LOADING RING DETECTOR

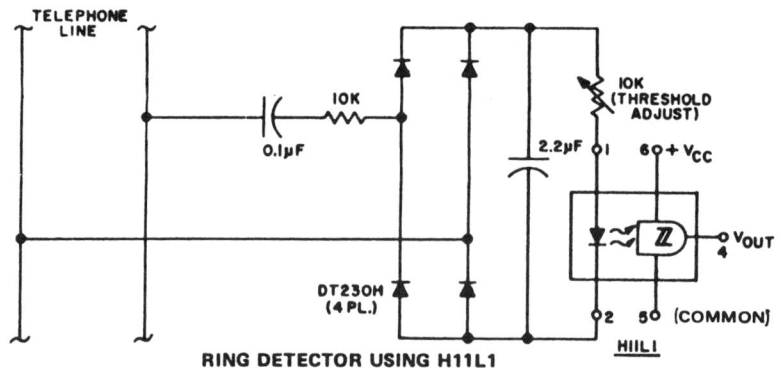

RING DETECTOR USING H11L1

GE

Fig. 30-6

RING DETECTOR *(Cont.)*

This circuit detects the 20 Hz, ≈ 86-V rms ring signal on telephone lines and initiates action in an electrically isolated circuit. Typical applications would include automatic answering equipment, and interconnect/interface and key systems. The circuits illustrated are *bare bones* circuits, designed to illustrate concepts. They might not eliminate the ac/dc ring differentiation, 60-Hz noise rejection, dial tap rejection, and other effects that must be considered in field application. The first ring detector is the simplest and provides about 1-mA signal for a 7-mA line loading for $^1/_{10}$ sec after the start of the ring signal. The time delay capacitor provides a degree of dial tap and click suppression, as well as filtering out the zero crossing of the 20-Hz wave. This circuit provides the basis for a simple example, a ring extender that operates lamps and buzzers from the 120-V 60-Hz power line, while maintaining positive isolation between the telephone line and the power line. Use of the isolated tab triac simplifies heatsinking by removing the constraint of isolating the triac heatsink from the chassis. Lower line current loading is required in many ring detector applications. This can be provided by using the H11BX522 photo-Darlington optocoupler, which is specified to provide a 1-mA output from a 0.5-mA input through the −25°C to +50°C temperature range.

The next circuit allows ring detection down to a 40-V rms ring signal, while providing 60-Hz rejection to about 20-V rms. Zero-crossing filtering can be accomplished either at the input bridge rectifier or at the output. Dependable ring detection demands that the circuit responds only to ring signals, rejecting spurious noise of similar amplitude, such as dialing transients. The configuration shown relies on the fact that ring signals are composed of continuous frequency bursts, whereas dialing transients are much lower in repetition rate. The dc bridge-filter combination at the H11L input has a time constant; it cannot react to widely spaced dialing transients, but will detect the presence of relatively long duration bursts, causing the H11L to activate the downstream interconnect circuits at a precisely defined threshold.

TELEPHONE OFF-HOOK INDICATOR

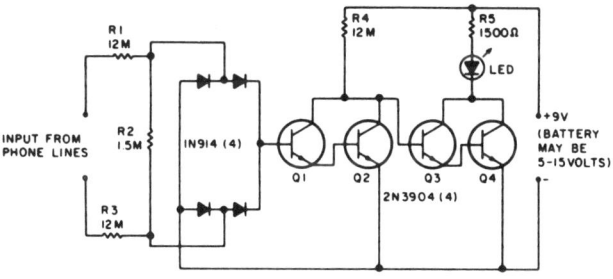

73 AMATEUR RADIO   *Fig. 30-7*

The LED flickers when the phone is ringing or being dialed. It glows steadily when the phone is off the hook.

FREQUENCY- AND VOLUME-CONTROLLED TELEPHONE RING DETECTOR

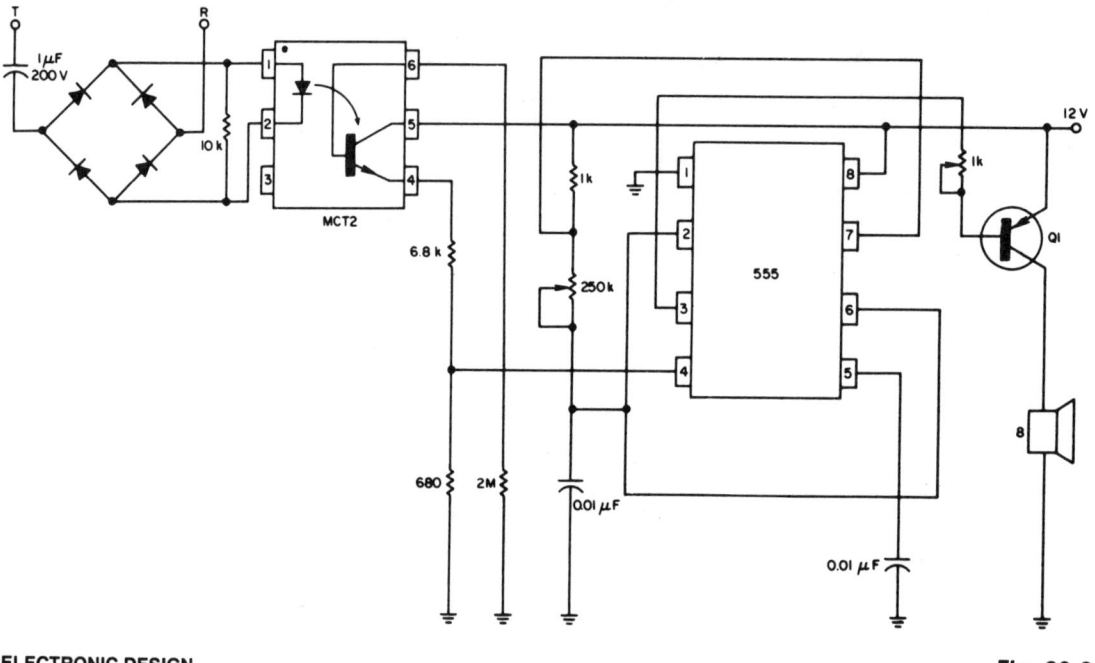

ELECTRONIC DESIGN

Fig. 30-8

With the 555 timer connected as a multivibrator and as an optoisolator, a remote speaker can be driven.

31

Tilt/Level Detectors

The sources of the following circuits are contained in the Sources section, which begins on page 214. The figure number in the box of each circuit correlates to the source entry in the Sources section.

Sense-of-Slope Tiltmeter
Differential-Capacitance Measurement Circuit
Ultra-Simple Level

SENSE-OF-SLOPE TILTMETER

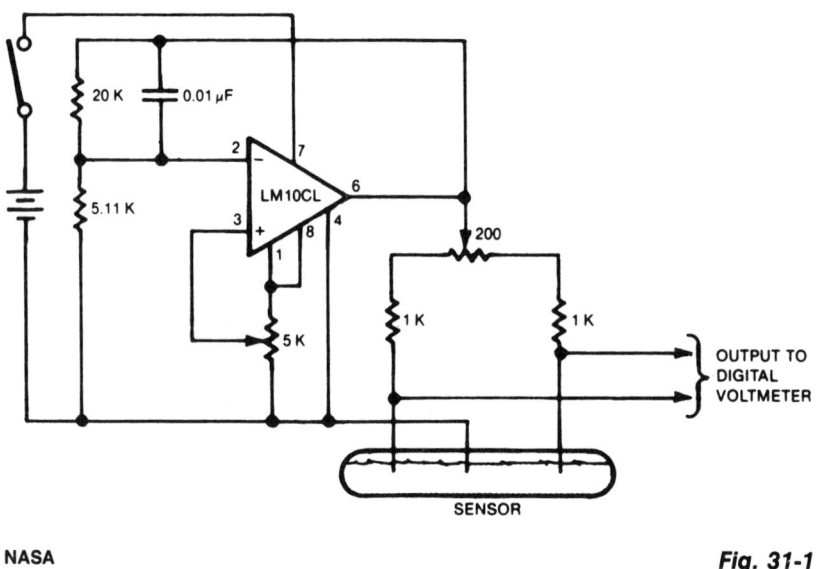

NASA

Fig. 31-1

Electrodes are immersed in an electrolyte that remains level while the sensor follows the tilt of the body on which it is placed; more of one outer electrode and less of the other are immersed and their resistances fall or rise, respectively. The resistance change causes a change in the output voltage of the bridge circuit. The sensor forms the two lower legs of the bridge, and two 1000-Ω metal-film resistors and a 200-Ω ceremet balance potentiometer form the two upper legs. In preparation for use, the bridge is balanced by adjusting the balance potentiometer so that the bridge output voltage is zero when the sensor is level. The bridge input voltage (dc excitation) is adjusted to provide about 10-mV output per degree of slope; the polarity indicates the sense of the slope. This scaling factor allows the multimeter to read directly in degrees if the user makes a mental shift of the meter decimal point. The scaling-factor calibration is done at several angles to determine the curve of output voltage versus angle.

DIFFERENTIAL-CAPACITANCE MEASUREMENT CIRCUIT

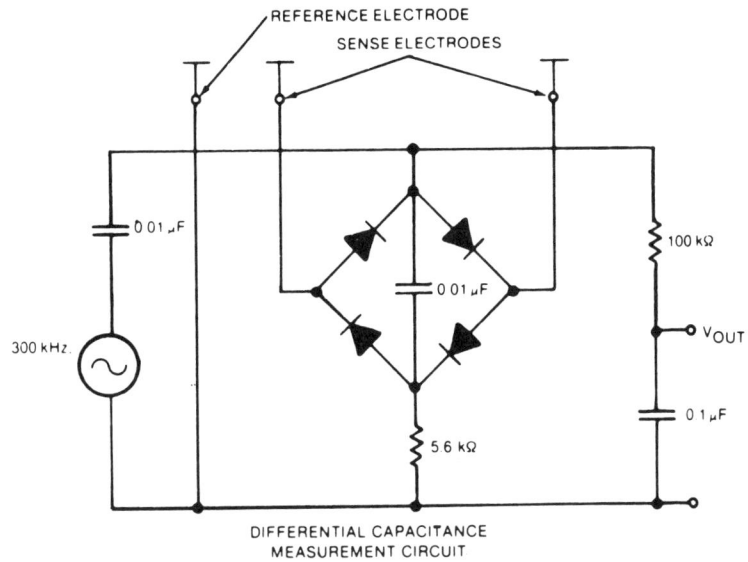

DIFFERENTIAL CAPACITANCE
MEASUREMENT CIRCUIT

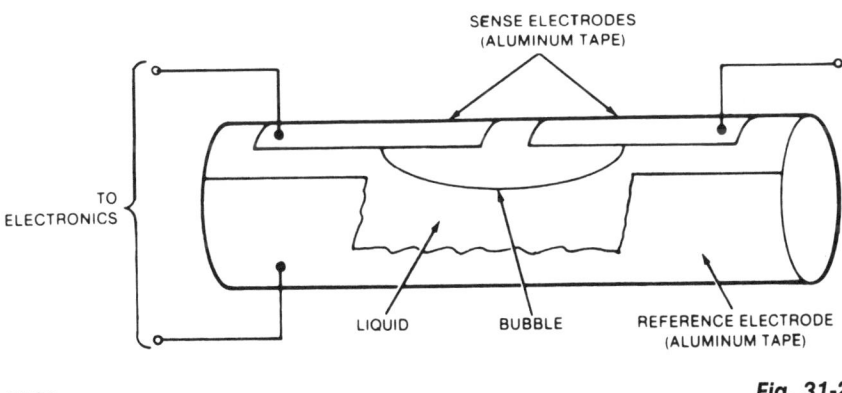

NASA

Fig. 31-2

A bubble vial with external aluminum-foil electrodes is the sensing element for a simple indicating tiltmeter. To measure bubble displacement, a bridge circuit detects the difference in capacitance between the two sensing electrodes and the reference electrode. Using this circuit, a tiltmeter level vial with 2-mm deflection for 5 arc-seconds of tilt easily resolves 0.05 arc-second. The four diodes are CA3039 (or equivalent).

ULTRA-SIMPLE LEVEL

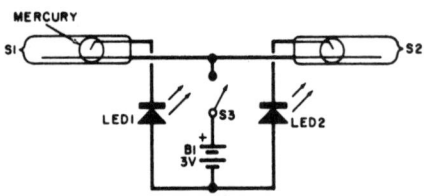

POPULAR ELECTRONICS   *Fig. 31-3*

This electronic level uses two LED indicators instead of an air bubble. If the surface is tilted to the right, one LED lights; if it's tilted to the left, the other LED lights. When the surface is level, both LEDs light. It uses two unidirectional mercury switches, S1 and S2. The unidirectional mercury switch has one long electrode and one short, angled electrode. The pool of mercury "rides" on the long electrode and makes contact between the two electrodes if the unit is held in a horizontal position.

32

Touch Detectors

The sources of the following circuits are contained in the Sources section, which begins on page 214. The figure number in the box of each circuit correlates to the source entry in the Sources section.

TOUCH ON/OFF SWITCH

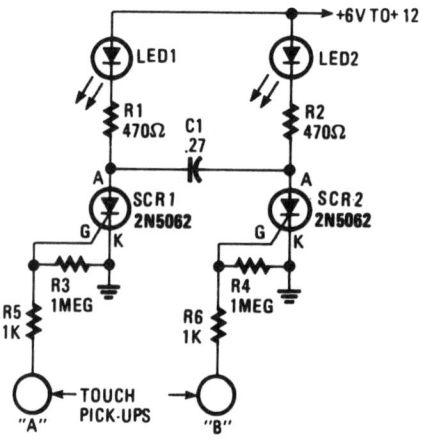

HANDS-ON ELECTRONICS   *Fig. 32-1*

Two sensitive-gate SCRs are interconnected so that when one of the devices is turned on, the other (if on) is forced off. That toggling effect gives an on/off circuit condition for each of the LEDs in the SCR-anode circuits. To turn LED1 on and LED2 off, simply touch the A terminal, and to turn LED1 off and LED2 on, the B pickup must be touched. It is possible to simultaneously touch both terminals, and cause both SCRs to turn on together. To reset the circuit to the normal one-on/one-off condition, momentarily interrupt the circuit's dc power source. Additional circuitry can be connected to the anode circuit of either or both SCRs to be controlled by the on/off function of the touch switch.

TOUCH SWITCH I

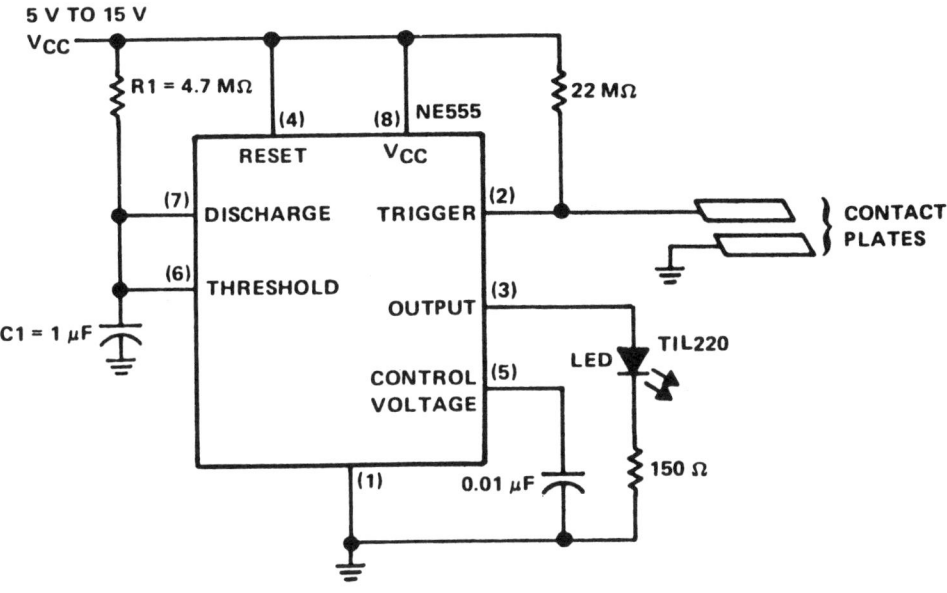

TEXAS INSTRUMENTS

Fig. 32-2

The circuit is basically a NE555 monostable; the only major difference is its method of triggering. The trigger input is biased to a high value by the 22-MΩ resistor. When the contact plates are touched, the skin resistance of the operator will lower the overall impedance from pin 2 to ground. This action will reduce the voltage at the trigger input to below the $1/3$ V_{CC} trigger threshold and the timer will start. The output pulse width will be $T = 1.1R_1C_1$, in this circuit about 5 seconds. A relay connected from pin 3 to ground, instead of to the LED and resistor, could be used to perform a switching function.

TOUCHOMATIC

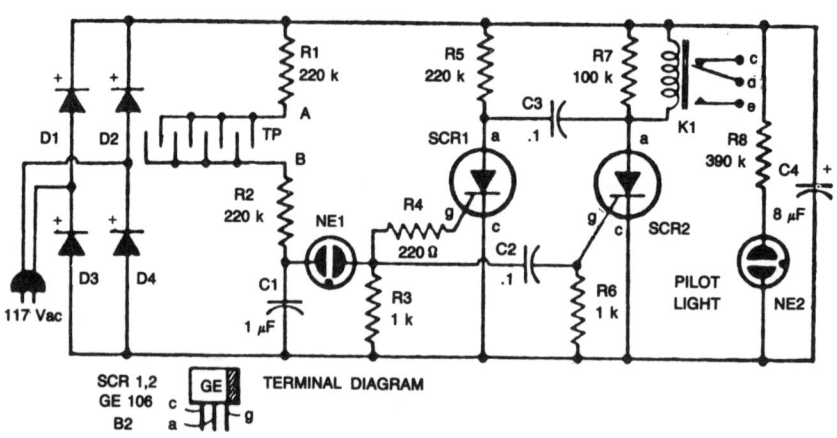

TAB BOOKS

Fig. 32-3

When someone touches the touchplate (TP), the resistance of his or her finger across points A and B is added in series to the combination of R_1 and R_2, and the capacitor C2 begins to charge. When the voltage across C1 is finally sufficient to fire NE1, C1 will begin to discharge. When NE1 fires, it produces a short between its terminals. Because R3 is connected across C1, they are effectively in series after NE1 fires. A voltage spike will then be passed by C2 and this will act as a positive triggering pulse. The pulse is fed to both SCR gates: SCR2 conducts, and thereby closes relay K1. With a finger no longer on the touchplate, no more pulses are forthcoming because the C1 charge path is open. The next contact with the touchplate will produce a pulse, which triggers SCR1. SCR2 is now off by capacitor C3, which was charged by current passing through R6 and SCR2. The firing of SCR1 in this way places a negative voltage across SCR2, which momentarily drops the relay current to a point below the holding current value of SCR2 (holding current is the minimum current an SCR requires to remain in a conducting state once its gate voltage is removed). With SCR2 turned off, the relay will open and SCR1 will turn off as a result of the large resistance in series with its anode. Starved in this way, SCR1 turns off because of a forced lack of holding current.

NEGATIVE-TRIGGERED TOUCH CIRCUIT

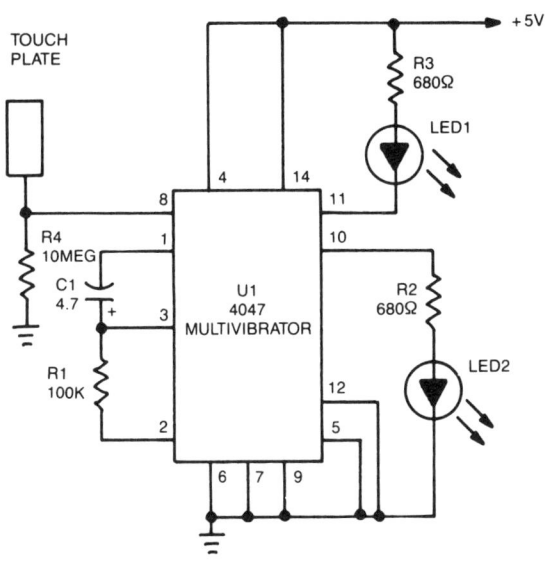

The 4047 is configured as a monostable multivibrator circuit or one shot that is set up to trigger on a negative-transition of the signal applied to its pin 6 input. The multivibrator's on time is determined by the values of R1 and C1. Although R1 is shown to be a 100-kΩ resistor, its value can be between 10 kΩ and 1 MΩ. Capacitor C1 can be a nonpolarized capacitor with any practical value above 100 pF. By making R4's value extremely high, the circuit can be used as a touch-triggered one-shot multivibrator. If the value of R4 is reduced to a much lower value, such as 10 kΩ, the circuit can be triggered with a negative pulse through a 0.1-μF capacitor connected to pin 6. With a 100-kΩ resistor for R1, and a 4.7-μF electrolytic capacitor for C1, the circuit's on time is about 0.6 second. When R1 is increased to 470 kΩ, the on time of the circuit is increased to over 6 seconds.

POPULAR ELECTRONICS   *Fig. 32-4*

POSITIVE-TRIGGERED TOUCH CIRCUIT

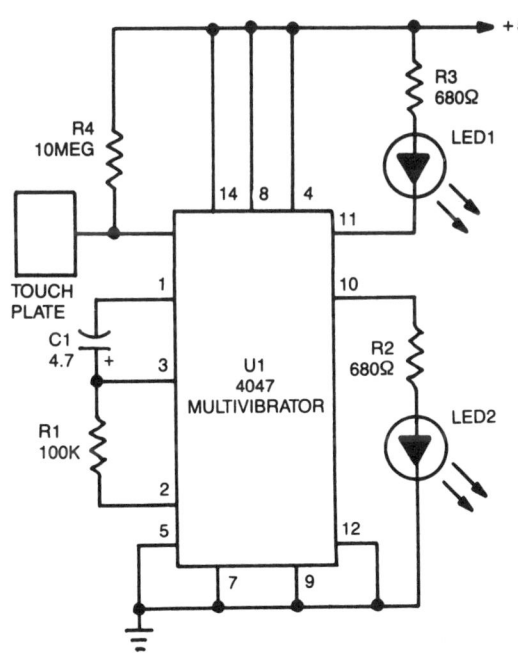

LED1 and LED2 indicators turn on and remain on, each time the circuit is triggered. During the timing cycle, U1's Q output at pin 10 becomes positive when the Q̄ output at pin 11 becomes negative. The two LEDs can be removed and the Q and Q̄ outputs at pins 10 and 11, respectively, can be used to trigger some other circuit.

POPULAR ELECTRONICS   *Fig. 32-5*

DIGITAL TOUCH ON/OFF SWITCH

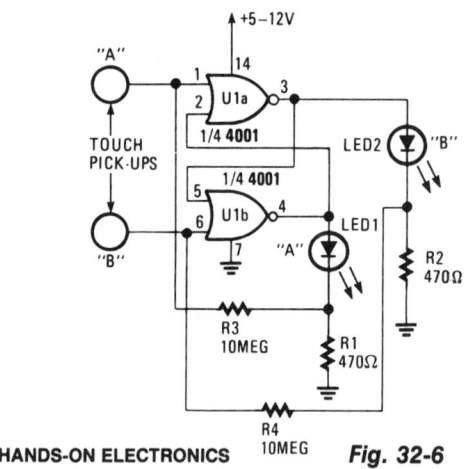

Only one LED can be on when the circuit is at rest. Which LED is illuminated is determined by the touch pickup that last had human contact. Pickup terminal A controls the on condition of LED1, and terminal B controls the on condition of LED2. A 4001 quad two-input NOR gate is connected in any anti-bounce latching circuit that is activated by touching a pickup.

HANDS-ON ELECTRONICS   *Fig. 32-6*

TWO-TERMINAL TOUCH SWITCH

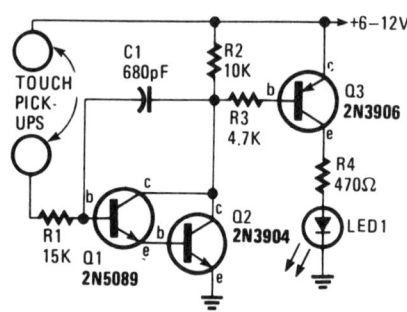

HANDS-ON ELECTRONICS   *Fig. 32-7*

This circuit requires the bridging of two circuits to activate the electronic switch. That circuit does not require a 60-Hz field to operate and can be battery or ac powered. The two-pickup terminals can be made from most any clean metal; they should be about the size of a penny. The input circuitry of the two-terminal touch switch is a high-gain Darlington amplifier that multiplies the small bridging current to a value of sufficient magnitude to turn on Q3, supplying power to LED1. If a quick on and off switching time is desired, C_1 should be very small; if a long on-time period is required, C_1 can be increased.

TOUCH ON/OFF ELECTRONIC SWITCH

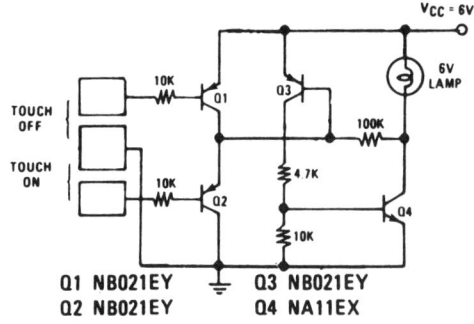

Transistors Q1 and Q2 control latch Q3 and Q4 to switch on the lamp. A high resistance from touching the electrode biases Q1 or Q2 on, and sets or resets the latch.

NATIONAL SEMICONDUCTOR CORP.   *Fig. 32-8*

LINE-HUM TOUCH SWITCH

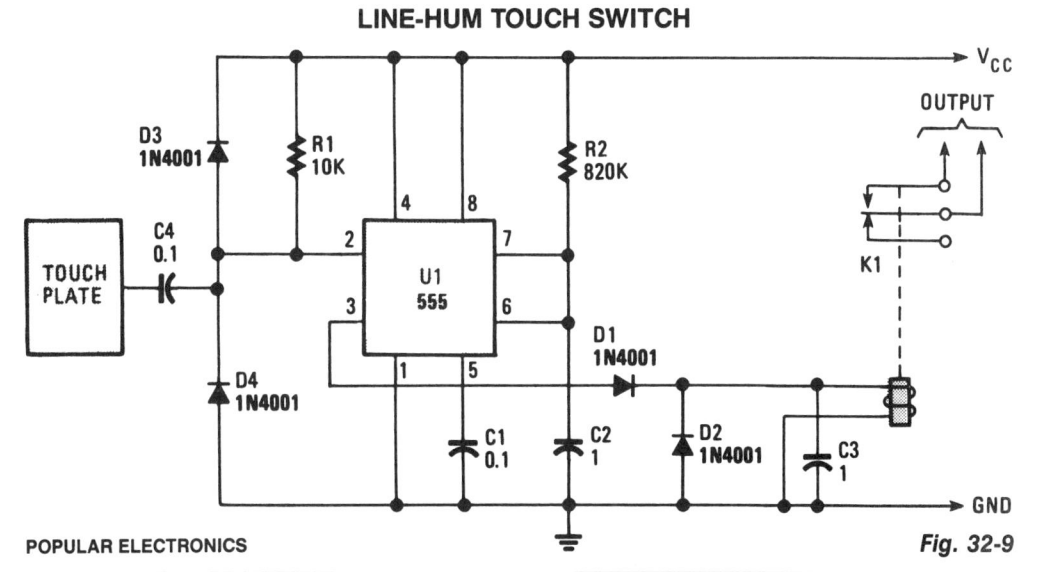

Fig. 32-9

The monostable period is set for about 1 second, as is the usual case. The induced line hum comes through C2, providing a continuous string of trigger pulses. The output becomes low for about 10 ms per second as the monostable times out and then retriggers. Diode D1 and capacitor C3 buffer the relay so that it doesn't chatter on those 10-ms pulses. Resistor R2 sets the sensitivity.

The relay energizes when the plate is touched and de-energizes, up to one second after the finger is removed. The delay is a function of when the monostable last retriggered.

TOUCH SWITCH II

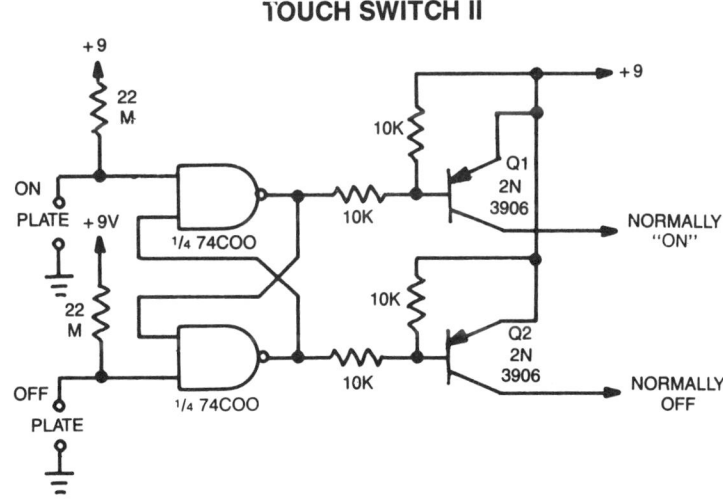

WILLIAM SHEETS

Fig. 32-10

When the plate is touched, the gate input becomes low, changing the state of the latch. Q1 and Q2 give alternate N-on/N-off outputs.

TOUCH SWITCH III

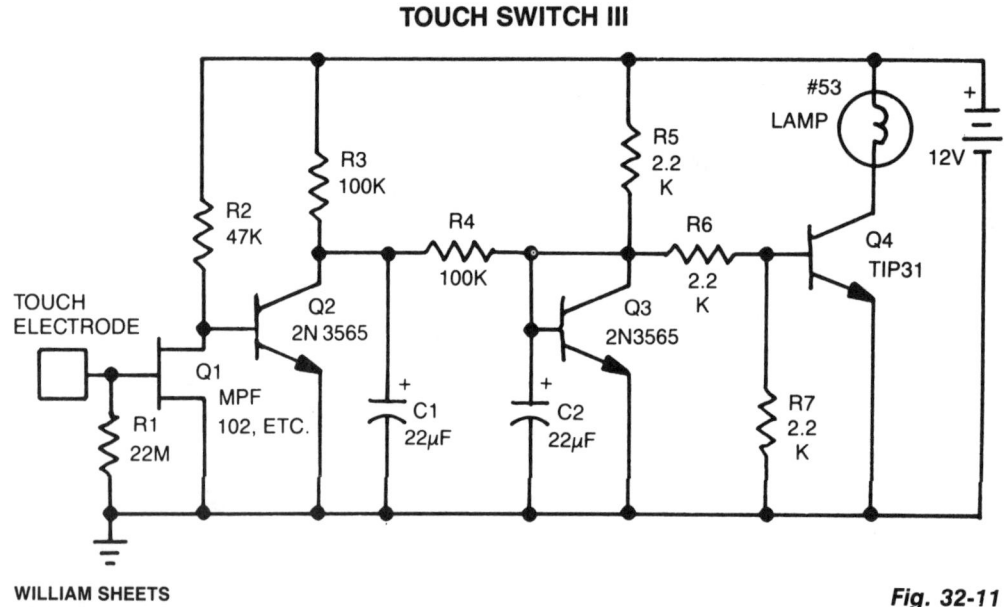

WILLIAM SHEETS

Fig. 32-11

This touch-actuated switch stays on as long as you keep your finger on the touch plate. R1 sets the input impedance to a high 22 MΩ. Q1 picks up stray signals coupled through your body to the touch plate and amplifies them to turn on Q2, which turns on lamp drivers Q3 and Q4. Lamp I1 is any small 12-V lamp, such as a No. 53 – 12 V 120 mA. R4 and C1 add a small amount of hysteresis (delay) to keep the light from constantly flickering. A relay can be used in place of the lamp.

TOUCH SWITCH IV

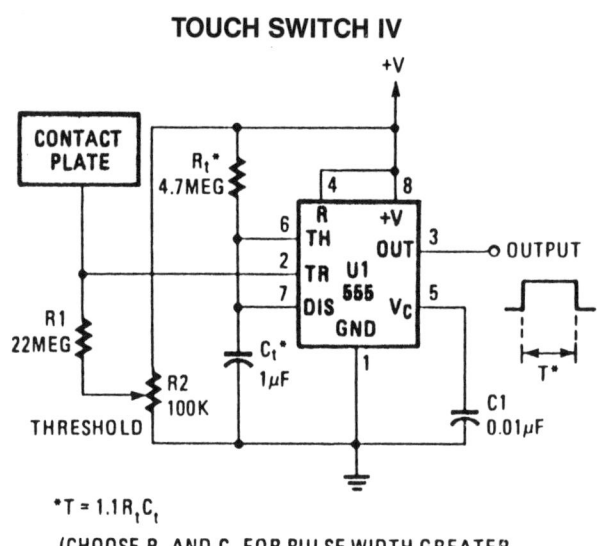

$^*T = 1.1 R_t C_t$

(CHOOSE R_t AND C_t FOR PULSE WIDTH GREATER THAN ANTICIPATED CONTACT TIME.)

HANDS-ON ELECTRONICS

Fig. 32-12

190

TOUCH CIRCUIT

SILICONIX

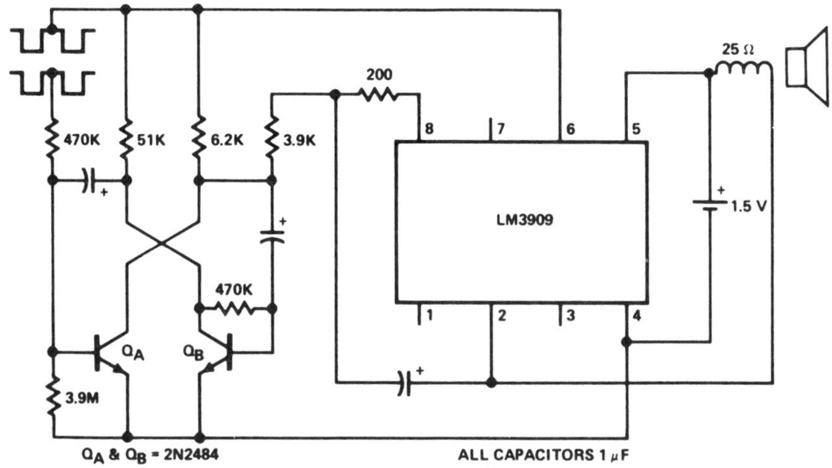

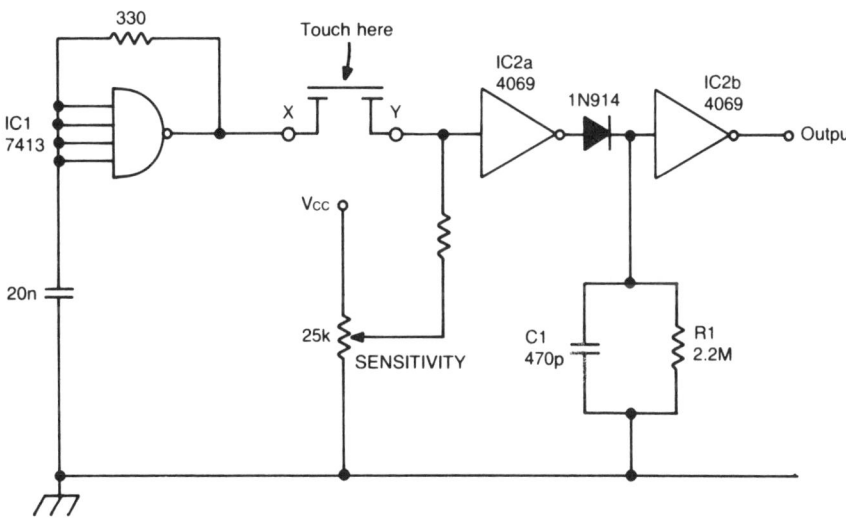

Fig. 32-13

CMOS TOUCH SWITCH

WIRELESS WORLD

Fig. 32-14

This touch switch does not rely on mains hum for switching. It can be used with battery-powered circuits. Schmitt trigger IC1 forms a 100-kHz oscillator and IC2a (which is biased into the linear region) amplifies the output and charges C1 via the diode. IC2b acts as a level detector. When the sensor is touched, the oscillator signal is severely attenuated, which causes C1 to discharge and IC2b to change state.

CAPACITANCE-OPERATED ALARM TO FOIL PURSE SNATCHERS

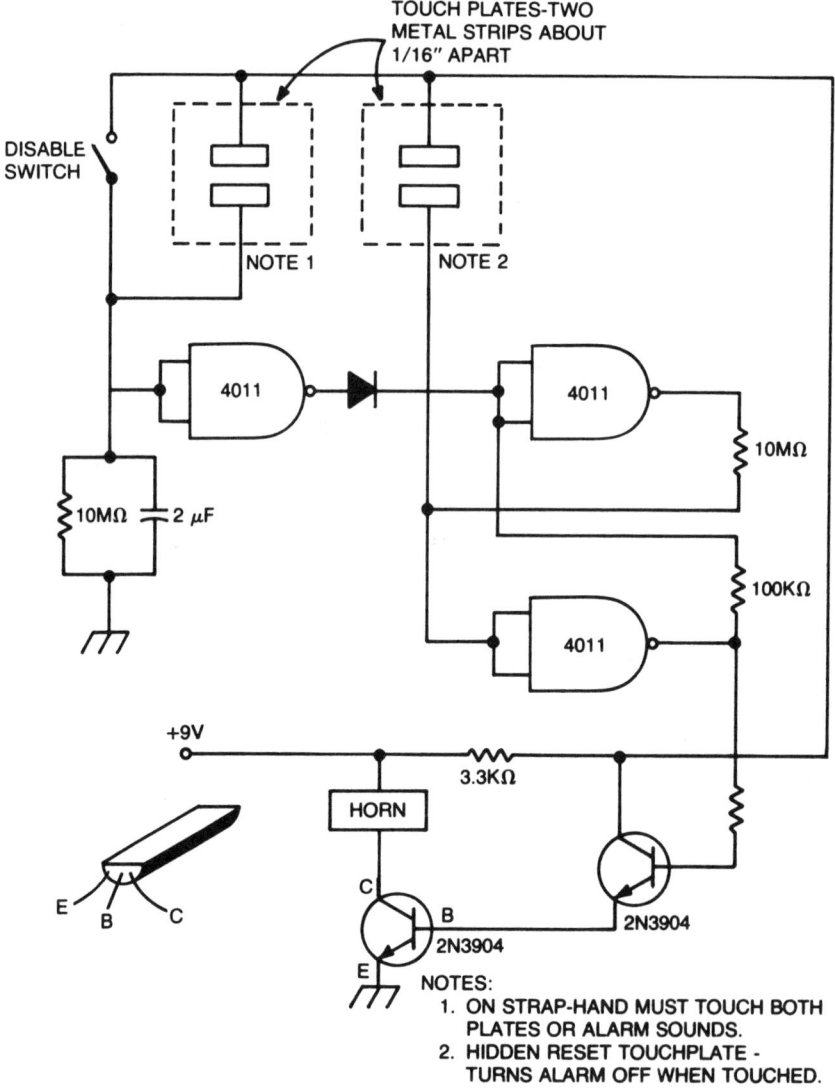

TOUCH PLATES-TWO
METAL STRIPS ABOUT
1/16" APART

DISABLE
SWITCH

NOTE 1   NOTE 2

4011

10MΩ ⏄ 2 µF

4011

10MΩ

100KΩ

4011

+9V

3.3KΩ

HORN

E B C

C

B
2N3904

E

2N3904

NOTES:
1. ON STRAP-HAND MUST TOUCH BOTH
   PLATES OR ALARM SOUNDS.
2. HIDDEN RESET TOUCHPLATE -
   TURNS ALARM OFF WHEN TOUCHED.

MODERN ELECTRONICS   *Fig. 32-15*

As long as touch plates (1) are touched together, the alarm is off. If not held for about 30 seconds, the alarm goes off. The circuit can be disabled with a switch or by touching the plates (2). The alarm is battery-operated and can sound a bicycle horn.

MOMENTARY-OPERATION TOUCH SWITCH

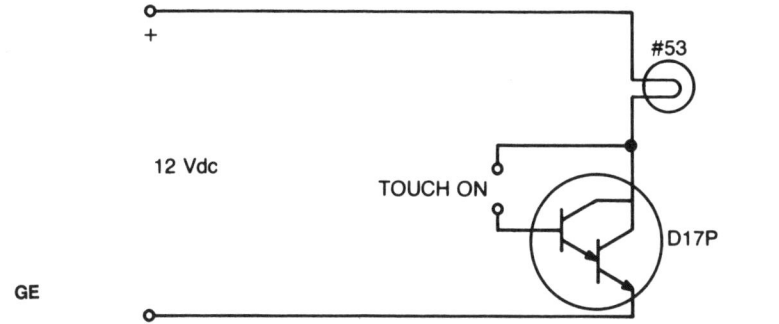

+

12 Vdc

TOUCH ON

#53

D17P

GE

Fig. 32-16

TOUCH-TRIGGERED BISTABLE

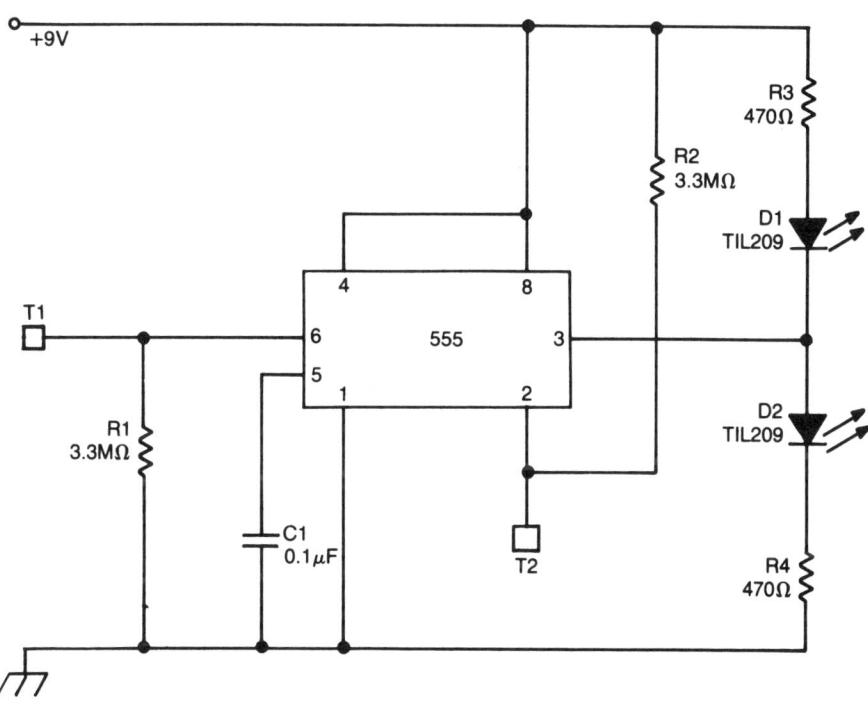

ELECTRONICS TODAY INTERNATIONAL

Fig. 32-17

This circuit uses a 555 timer in the bistable mode. Touching T2 causes the output to go high; D2 conducts and D1 extinguishes. Touching T1 causes the output to go low; D1 conducts and D2 is cut off. The output from pin 3 can also be used to operate other circuits (e.g., a triac-controlled lamp). In this case, the LEDs are useful for finding the touch terminals in the dark. C1 is not absolutely necessary, but it helps to prevent triggering from spurious pulses.

LOW-CURRENT TOUCH SWITCH

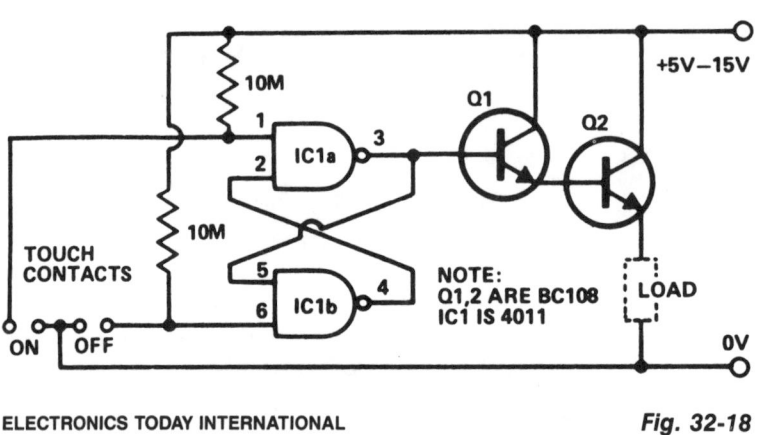

Fig. 32-18

Touching the on contacts with your finger brings pin 3 high, turning on the Darlington pair and supplying power to the load (transistor radio, etc). Q1 must be a high-gain transistor, and Q2 is chosen for the current required by the load circuit.

CAPACITANCE-SWITCHED LIGHT

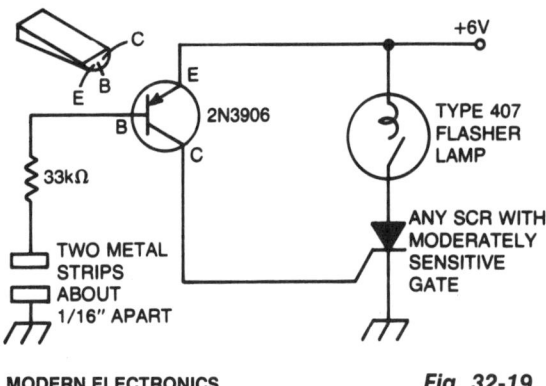

Fig. 32-19

The battery-powered light turns on easily, stays on for just a few seconds, then turns off again. The circuit is triggered when you place your finger across the gap between two strips of metal, about 1/16 inch apart. Enough current will flow through your finger to trigger the SCR after being amplified by the 2N3906. Once the SCR is fired, current will flow through the bulb until its internal bimetal switch turns it off. Once that happens, the SCR will return to its nonconducting state.

CAPACITANCE-OPERATED BATTERY-POWERED LIGHT

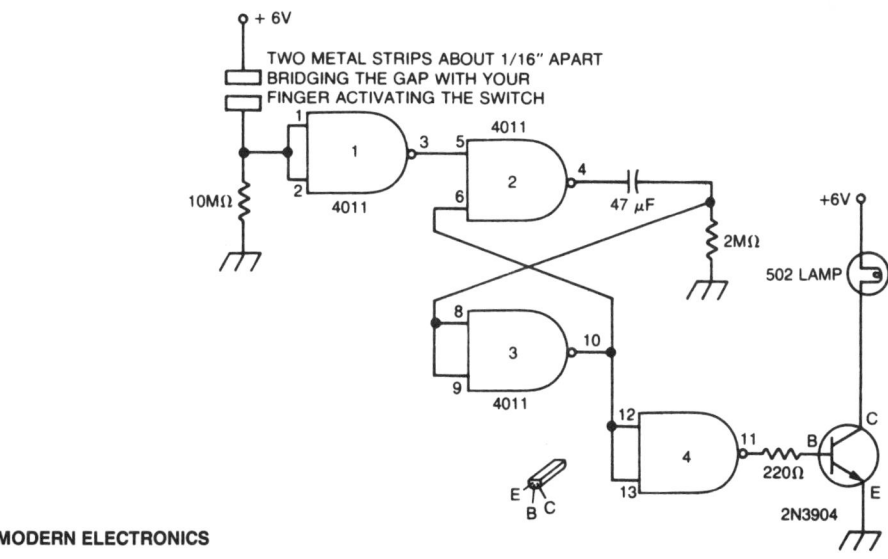

MODERN ELECTRONICS

Fig. 32-20

Touch the plate and the light will go on and remain on for a time determined by the time constant of the 47-μF capacitor and the 2-MΩ resistor.

TOUCH-SENSITIVE SWITCH

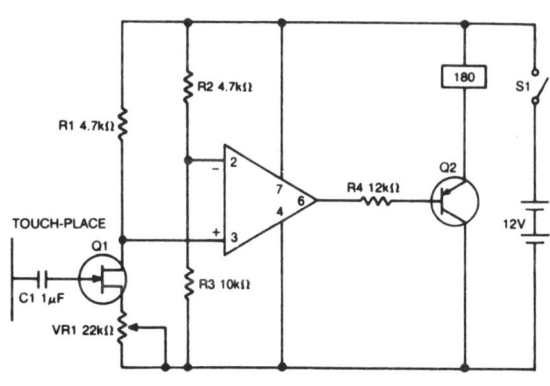

ELECTRONICS TODAY INTERNATIONAL **Fig. 32-21**

A high-impedance input is provided by Q1, a general-purpose field-effect transistor. A 741 op amp is used as a sensitive voltage-level switch, which in turn operates Q2, a medium-current pnp bipolar transistor. Thereby, the relay is energized and it can be used to control equipment, alarms, etc.

LATCHING DOUBLE-BUTTON TOUCH SWITCH

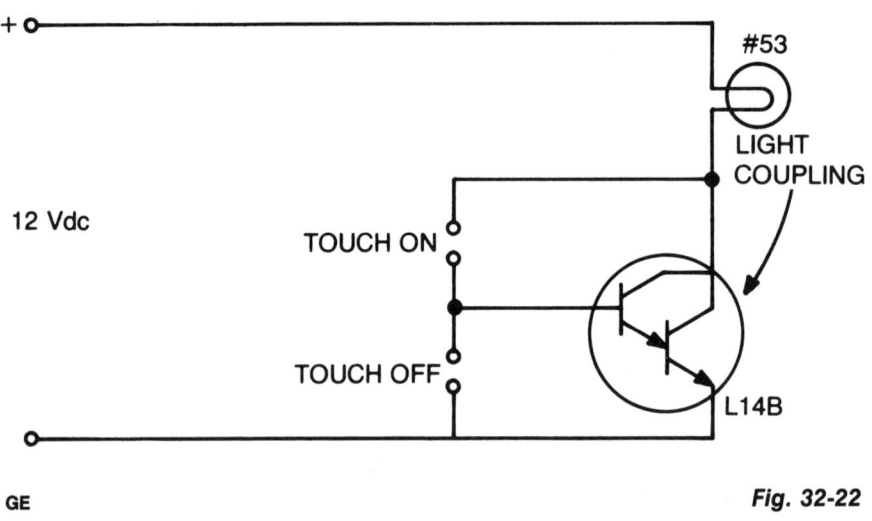

Fig. 32-22

FINGER-TOUCH OR CONTACT SWITCH

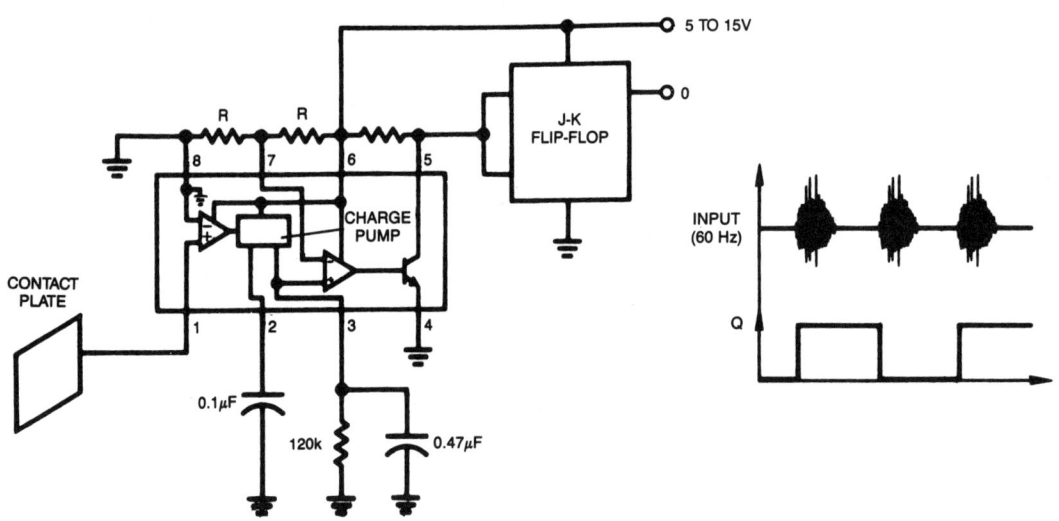

Fig. 32-23

33

Voice/Sound Detectors

The sources of the following circuits are contained in the Sources section, which begins on page 214. The figure number in the box of each circuit correlates to the source entry in the Sources section.

VOICE-ACTIVATED SWITCH AND AMPLIFIER

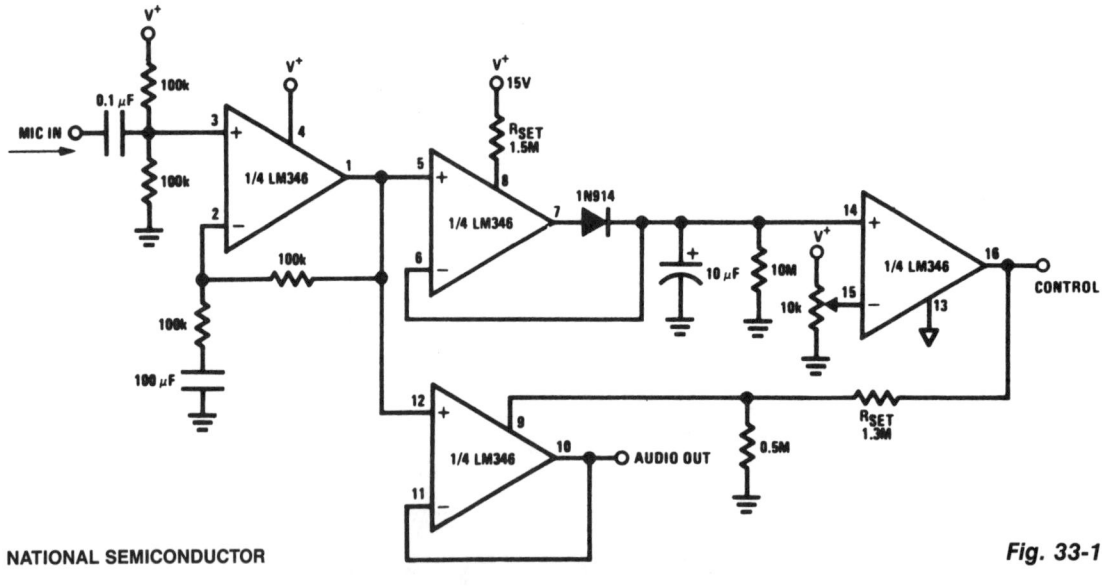

NATIONAL SEMICONDUCTOR

Fig. 33-1

AUDIO-OPERATED RELAY

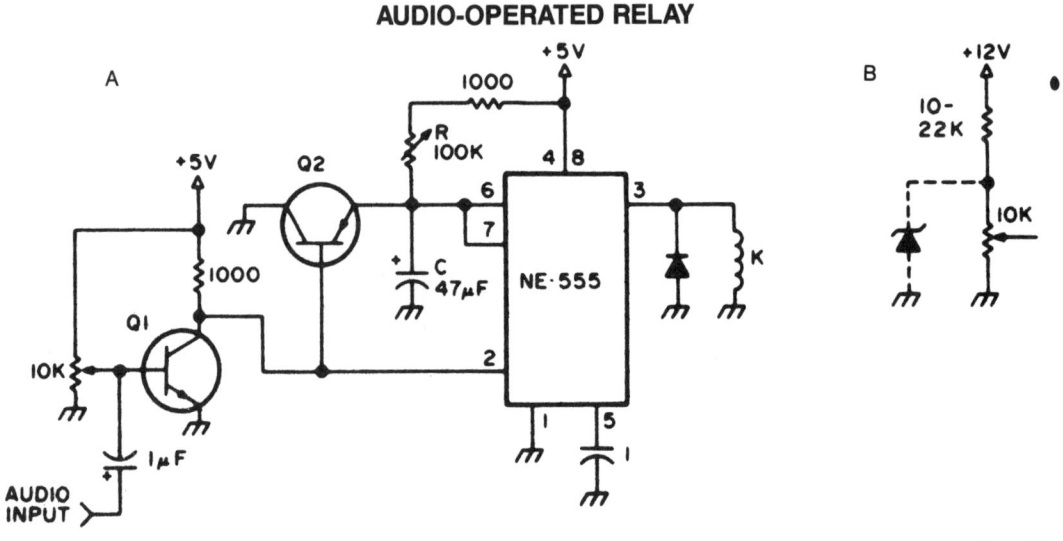

73 AMATEUR RADIO

Fig. 33-2

Q1 and Q2 are general-purpose transistors. The 10-kΩ input pot is adjusted to a point just short of where Q1 turns on (as indicated by K pulling in). K is any 5-V reed relay. With the values shown for R (100 kΩ) and C (47 μF), timing values from 0.05 to slightly over 5 seconds can be achieved. B shows the addition of a 22-kΩ series resistor to the 10-kΩ input pot if a 12-V supply is used. A suitable 12-V reed relay must be used at K.

SOUND-ACTIVATED RELAY

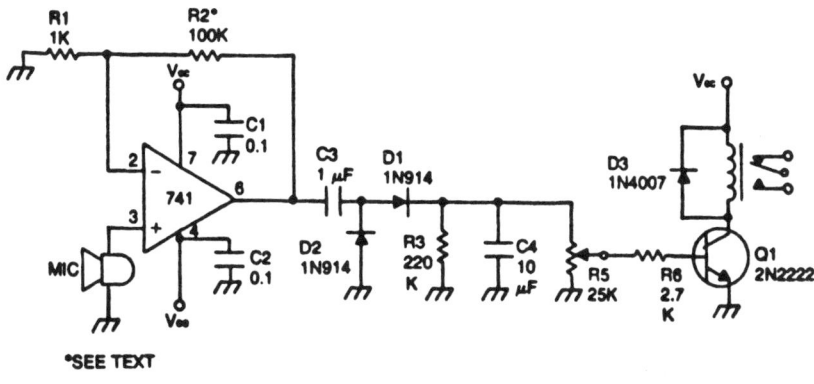

Fig. 33-3

The device remains dormant (in an off condition) until some sound causes it to turn on. The input stage is a 741 operational amplifier that is connected as a noninverting follower audio amplifier. Gain is approximately 100. To increase gain, raise the value of R_2. The amplified signal is rectified and filtered to a dc level by R4. Then, R5 is set to the audio level desired to activate the relay.

SOUND-OPERATED TWO-WAY SWITCH

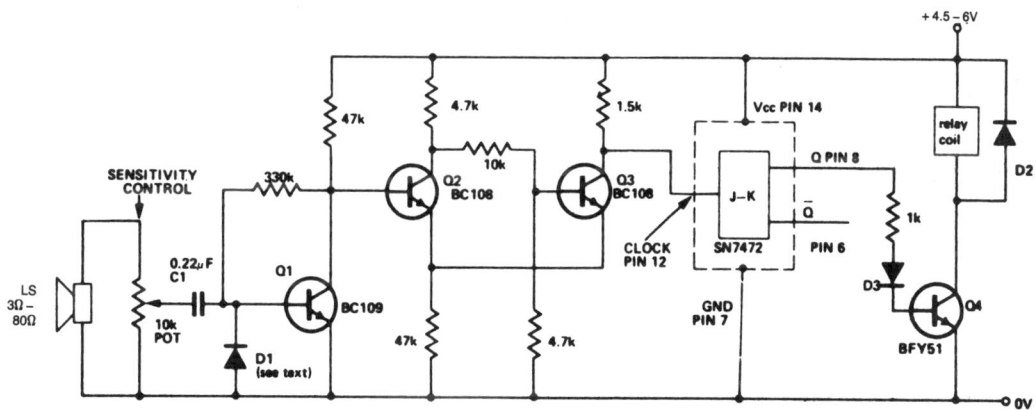

Fig. 33-4

This circuit operates a relay each time a sound of sufficient intensity is made; thus, one clap of the hands will switch it one way, a second clap will revert the circuit to the original condition. Q2 and Q3 form a Schmitt trigger. The JK flip-flop is used as a bistable whose output changes state every time a pulse is applied to the clock input (pin 12). Q4 allows the output to drive a relay.

SOUND-MODULATED LIGHT SOURCE

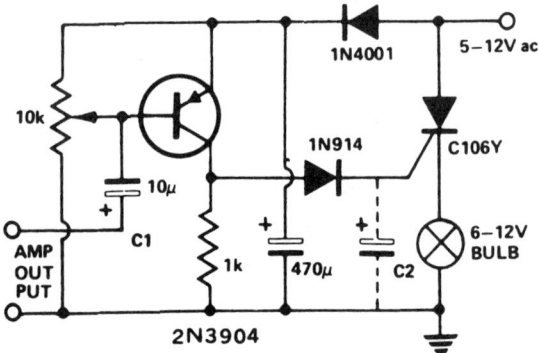

This circuit modulates a light beam with voice or music from the output of an amplifier. If the 10-kΩ pot is adjusted to slightly less than the V_{be} of the transistor, the circuit forms a peak detector. This drives the gate of the SCR, lighting the bulb whose brightness will vary as the sound level varies. C2 can be left out for a faster response.

ELECTRONICS TODAY INTERNATIONAL   *Fig. 33-5*

AUDIO-CONTROLLED LAMP

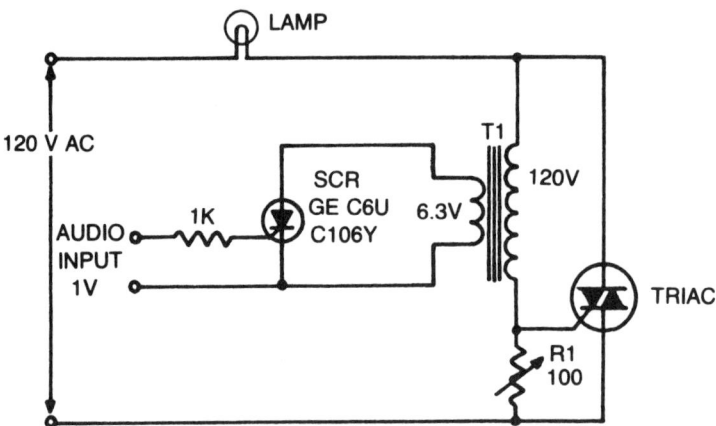

NOTE: T1 IS A 6.3V, 1A "FILAMENT" TRANSFORMER. ADJUST R1 FOR MAXIMUM RESISTANCE THAT WILL <u>NOT</u> TURN ON LAMP WITH ZERO INPUT.

GE   *Fig. 33-6*

This is an on/off control with an isolated, low-voltage input. Because the switching action is very rapid, compared to the response time of the lamp and the response of the eye, the effect produced with audio input is similar to a proportional-control circuit. If the input signal to the SCR consists of phase-controlled pulses, full-wave control of the lamp load is obtained.

SOUND-ACTIVATED SWITCH

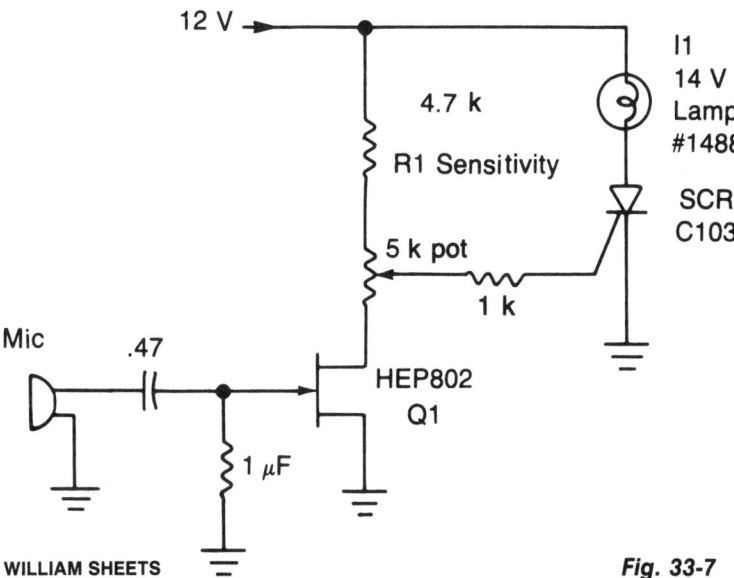

WILLIAM SHEETS

Fig. 33-7

The audio from Mic is amplified by Q1. Peaks of signal (adjusted by R1) greater than about 0.7 V trigger the SCR and light lamp I1.

SOUND-ACTIVATED AC SWITCH

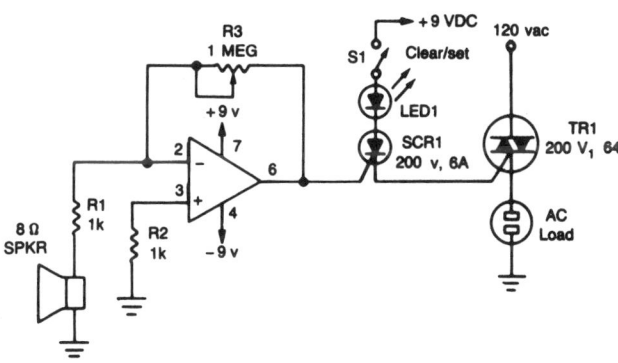

RADIO-ELECTRONICS

Fig. 33-8

The circuit uses a 741 op amp operating as an inverting amplifier to amplify the voltage produced by an 8-Ω speaker used to detect any sounds. The feedback resistor R3, a 1-MΩ potentiometer used to vary the gain of the amplifier determines the sensitivity of the circuit. When S1 is closed in the (SET) position and a sound is applied to the speaker, SCR1 is turned on. It will remain in conduction until the anode voltage is removed by opening S1, putting it in its RESET position (once an SCR is turned on, the gate or trigger has no control over the circuit). As long as the SCR conducts, the triac, TR1, will remain on and supply voltage to the load.

34

Window Detectors

The sources of the following circuits are contained in the Sources section, which begins on page 214. The figure number in the box of each circuit correlates to the source entry in the Sources section.

WINDOW DETECTOR I

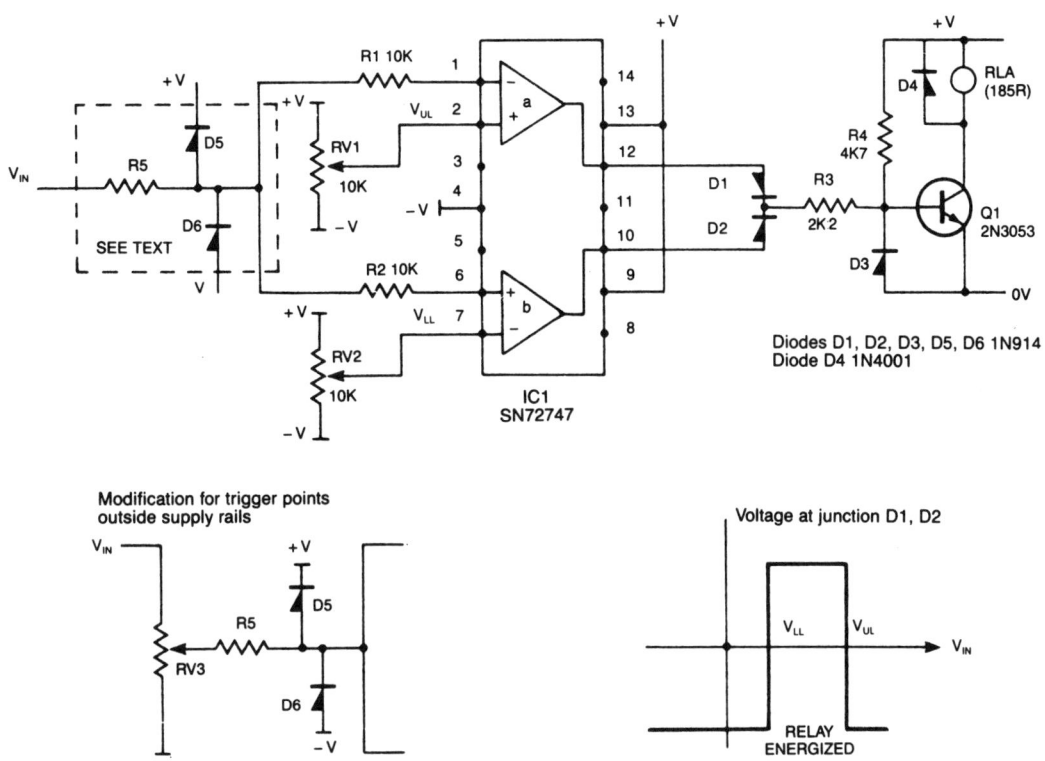

ELECTRONICS TODAY INTERNATIONAL

Fig. 34-1

This circuit de-energizes a normally energized relay if the input voltage goes above or below two individually set voltages. The transistor driving the relay is normally turned on by R4, so the relay is normally energized. If the cathode of D1 or D2 is taken negative, Q1 will turn off and the relay will de-energize. The IC is a 72747 dual op amp used without feedback, so the full gain of about 100 dB is available. The amplifier output will thus swing from full positive to full negative for a few mV change at the input. The relay is therefore only energized if V_{IN} is between V_{UL} and V_{LL}. The two limits can be set anywhere between the supply rails, but obviously V_{UL} must be more positive than V_{LL}. If V_{IN} can go outside the supply rails, D5, D6, and R5 should be added to prevent damage to IC1. If V_{UL} and V_{LL} are required to be outside the supply rails, V_{IN} can be reduced by RV3. The supplies can be any value, provided that the voltage across them is not more than 30 V.

WINDOW DETECTOR II

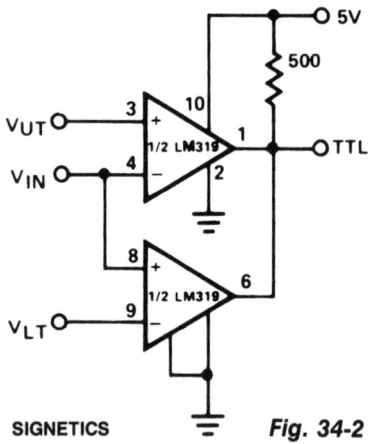

$$V_{OUT} = 5V \text{ for } V_{LT} < V_{IN} < V_{UT}$$
$$V_{OUT} = 0 \text{ for } V_{IN} < V_{LT} \text{ or } V_{IN} > V_{UT}$$

SIGNETICS · Fig. 34-2

WINDOW DETECTOR III

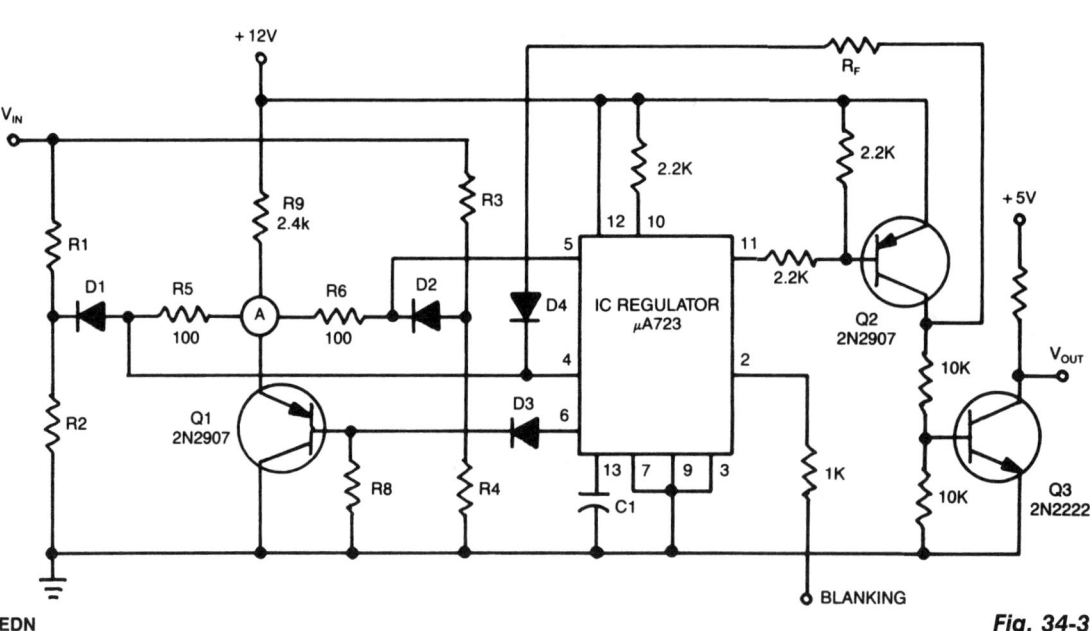

EDN · Fig. 34-3

The detector circuit compares the output voltage of two separate voltage dividers with a fixed reference voltage. The resultant absolute-error signal is amplified and converted to a logic signal that is TTL compatible.

MULTIPLE-APERTURE WINDOW DISCRIMINATOR

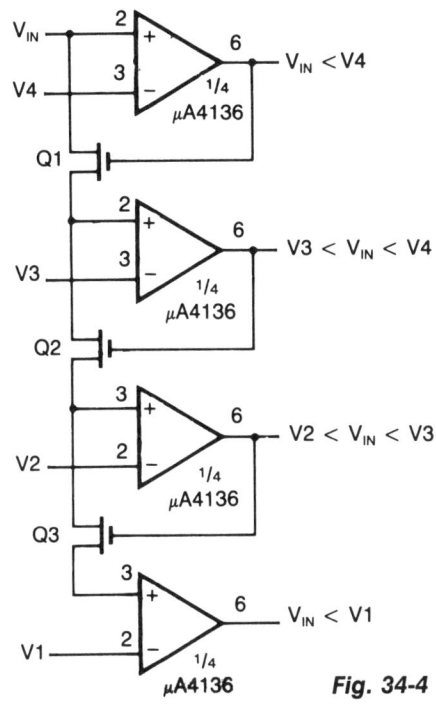

The circuit shown here uses μA4136 comparators and FETs Q1 through Q3.

Fig. 34-4

FAIRCHILD CAMERA AND INSTRUMENT CORP.

WINDOW COMPARATOR I

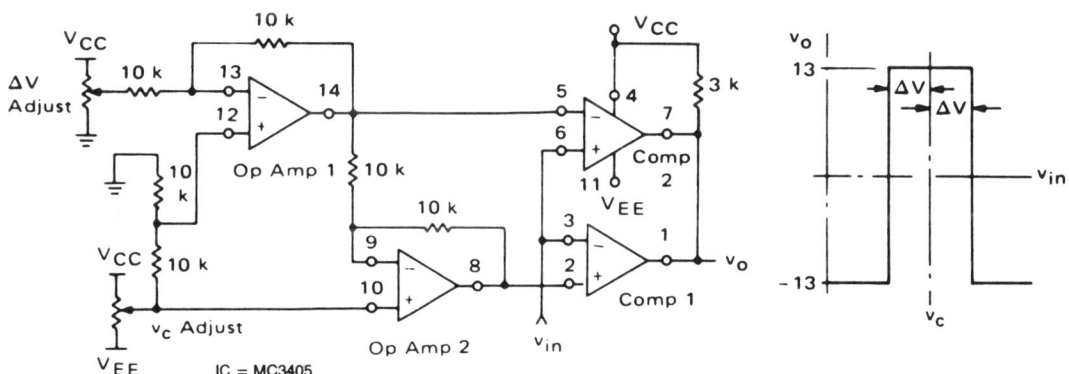

TEXAS INSTRUMENTS

Fig. 34-5

WINDOW COMPARATOR II

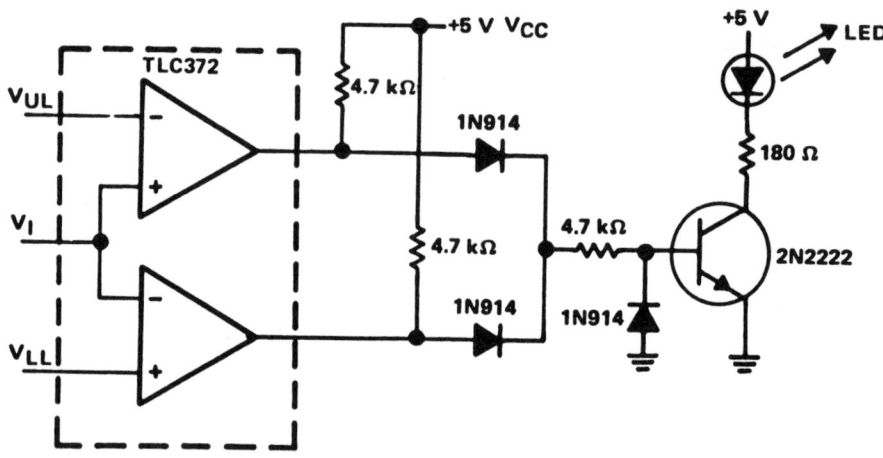

Window Comparator with LED Indicator

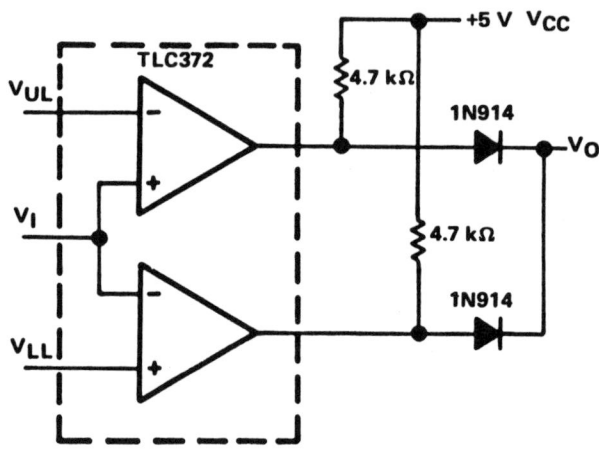

Basic Window Comparator

TEXAS INSTRUMENTS

Fig. 34-6

A window detector is a specialized comparator circuit designed to detect the presence of a voltage between two prescribed limits; that is, within a voltage *window*. This circuit is implemented by logically combining the outputs of two single-ended comparators by the 1N914 diodes. When the input voltage is between the upper limit, V_{UL}, and the lower limit, V_{LL}, the output voltage is zero; otherwise it equals a logic high level. The output of this circuit can be used to drive a logic gate, LED driver, or relay driver circuit. The circuit shows a 2N2222 npn transistor being driven by the window comparator. When the input voltage to the window comparator is outside the range set by the V_{UL} and V_{LL} inputs, the output changes to positive, which turns on the transistor and lights the LED indicator.

WINDOW DETECTOR IV

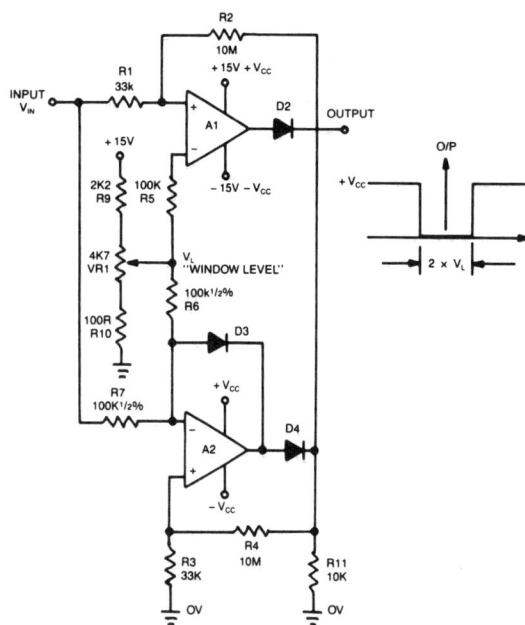

This novel window detector uses only two op amps. The width of the window can be changed by the 4.7-kΩ potentiometer.

ELECTRONIC ENGINEERING

Fig. 34-7

SIMPLE WINDOW DETECTOR

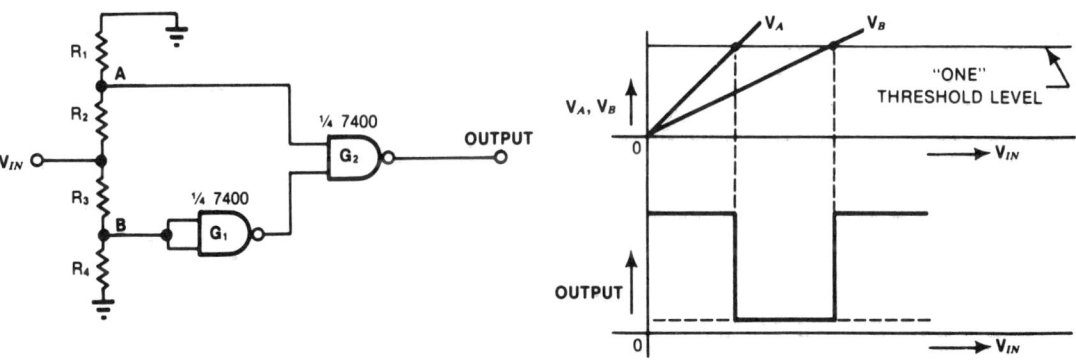

EDN

This simple window detector uses only half of a 7400 quad NAND gate, plus four resistors, chosen so that the voltage at point A exceeds the voltage at point B for any input voltage. With no input applied or when V_{IN} is at ground, the output of gate G1 is one; hence G2's output is also one. As the input voltage increases, V_A rises faster than V_B. When V_A reaches an acceptable one level, the circuit's output drops to zero. As the input continues to increase, V_B rises to an acceptable level, changing the output of G2 to one.

DIGITAL FREQUENCY WINDOW

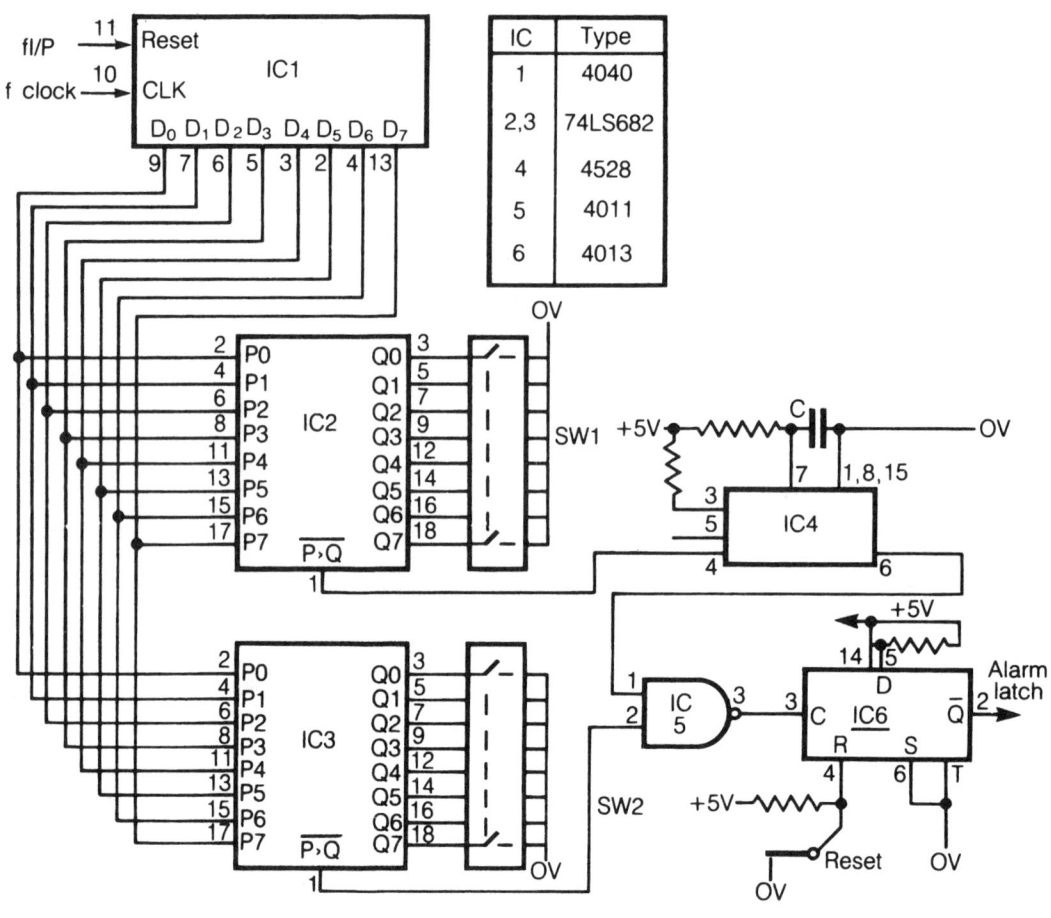

IC	Type
1	4040
2,3	74LS682
4	4528
5	4011
6	4013

ELECTRONIC ENGINEERING

Fig. 34-9

This circuit detects frequency variation above or below preset limits. IC1 is a binary counter clocked at F_{CLK}. The outputs are compared with switch preset values by IC2 and IC3. The input signal, which must be a positive-going pulse, is used to reset IC1. The *P greater than Q* output of the comparators is at logic 0 for input frequencies below the preset values. Above the preset count, a pulse train is output.

IC2, detects a low input by supplying the pulse train to a retriggerable monostable, IC4. When the input frequency falls below the preset value in SW1, the monostable is no longer triggered and its output falls to logic 0. IC3 detects the frequency high-state SW2, and outputs directly when this occurs. The outputs from both comparators can then be latched (as shown), using IC5 and IC6. The clock frequency is related to input and switch values: switch value = $F_{CLK/input}$. The time constant of IC4 is not critical, but must obviously exceed the maximum input pulse period.

WINDOW DETECTOR V

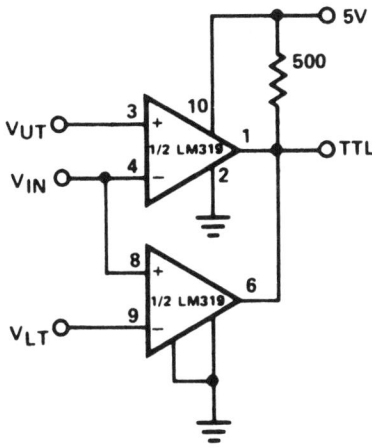

$V_{OUT} = 5V$ for $V_{LT} < V_{IN} < V_{UT}$
$V_{OUT} = 0$ for $V_{IN} < V_{LT}$ or $V_{IN} > V_{UT}$

SIGNETICS   *Fig. 34-10*

35

Zero-Crossing Detectors

The sources of the following circuits are contained in the Sources section, which begins on page 214. The figure number in the box of each circuit correlates to the source entry in the Sources section.

ZERO-CROSSING SWITCH

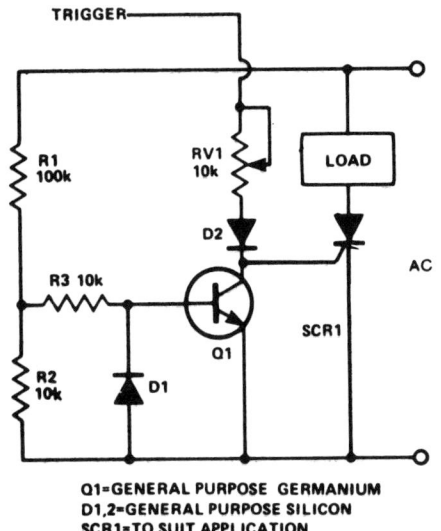

Q1=GENERAL PURPOSE GERMANIUM
D1,2=GENERAL PURPOSE SILICON
SCR1=TO SUIT APPLICATION

ELECTRONICS TODAY INTERNATIONAL   *Fig. 35-1*

When switching loads with the aid of a thyristor, a large amount of RFI can be generated unless some form of zero-crossing switch is used. The circuit shows a simple single-transistor zero-crossing switch. R1 and R2 act as a potential divider. The potential at their junction is about 10% of the ac voltage. This voltage level is fed, via R3, to the transistor's base. If the voltage at this point is above 0.2, the transistor will conduct, shunting any thyristor gate current to ground. When the line potential is less than about 2 V, it is possible to trigger the thyristor. The diode (D1) is to remove any negative potential that might cause reverse breakdown.

ZERO-CROSSING DETECTOR I

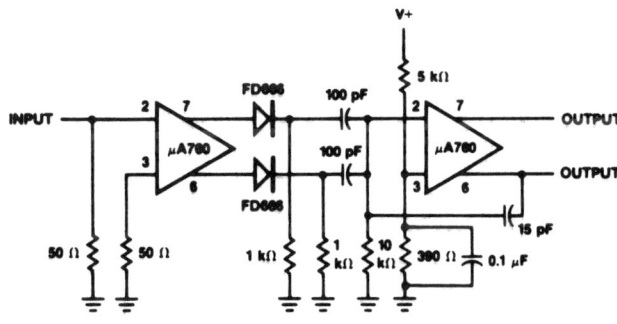

Total Delay = 30 ns
Input frequency = 300 Hz to 3 MHz
Minimum input voltage = 20 mVpk-pk

FAIRCHILD CAMERA & INSTRUMENT   *Fig. 35-2*

ZERO-CROSSING DETECTOR II

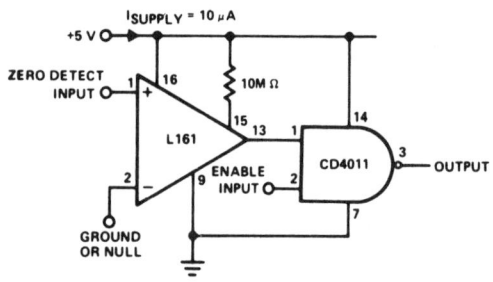

This detector is useful in sine-wave squaring circuits and A/D converters. The positive input can either be grounded or connected to a nulling voltage, which cancels input offsets and enables accuracy to within microvolts of ground. The CMOS output will switch to within a few millivolts of either rail for an input voltage change of less than 200 μV.

SILICONIX   Fig. 35-3

ZERO-CROSSING DETECTOR WITH TEMPERATURE SENSOR

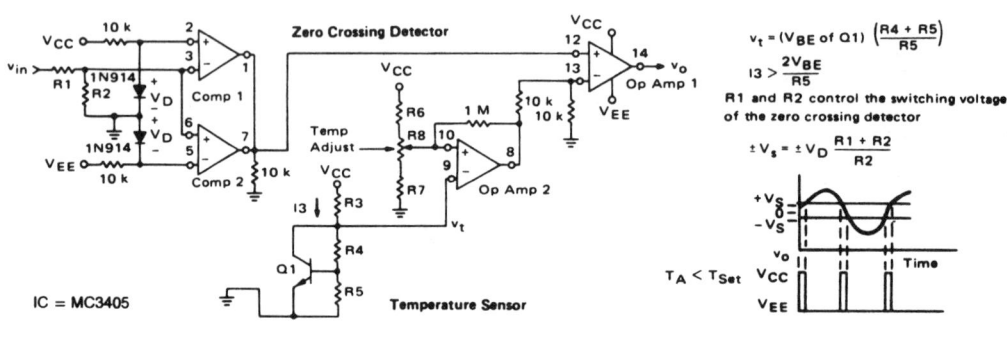

$$v_t = (V_{BE} \text{ of } Q1) \left(\frac{R4 + R5}{R5}\right)$$

$$I3 > \frac{2V_{BE}}{R5}$$

R1 and R2 control the switching voltage of the zero crossing detector

$$\pm V_s = \pm V_D \frac{R1 + R2}{R2}$$

MOTOROLA   Fig. 35-4

ZERO-CROSSING DETECTOR III

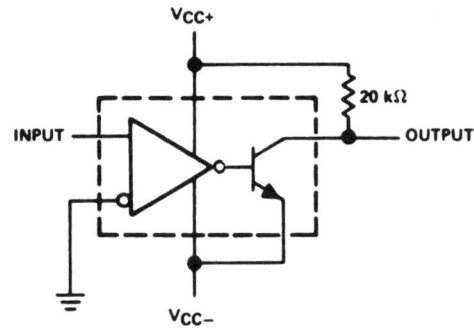

TEXAS INSTRUMENTS   Fig. 35-5

ZERO-CROSSING DETECTOR IV

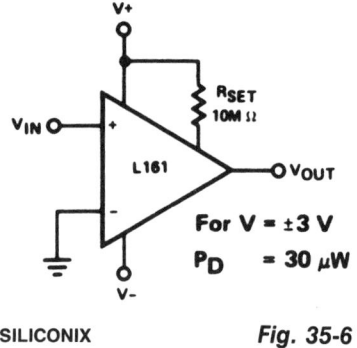

For V = ±3 V
P_D = 30 μW

SILICONIX   Fig. 35-6

ZERO-CROSSING DETECTOR V

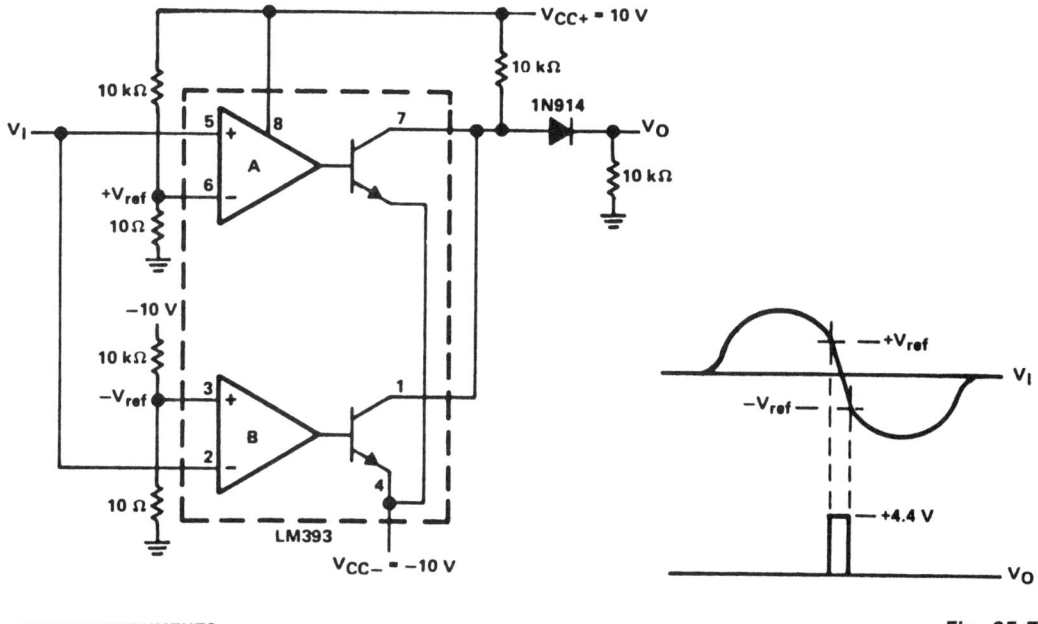

TEXAS INSTRUMENTS

Fig. 35-7

This zero-crossing detector uses a dual LM393 comparator, and easily controls hysteresis by the reference levels, which are set on the comparator inputs. The circuit illustrated is powered by ±10-V power supplies. The input signal can be an ac signal level up to +8 V. The output will be a positive-going pulse of about 4.4 V at the zero-crossover point. These parameters are compatible with TTL logic levels.

The input signal is simultaneously applied to the noninverting input of comparator A and the inverting input of comparator B. The inverting input of comparator A has a +10-mV reference, with respect to ground; the noninverting input of comparator B has a −10-mV reference, with respect to ground. As the input signal swings positive (greater than +10 mV), the output of comparator "A" will be low, while comparator "B" will have a high output. When the input signal swings negative (less than −10mV), the reverse is true. The result of the combined outputs will be low in either case. On the other hand, when the input signal is between the threshold points (±10 mV around zero crossover), the output of both comparators will be high. If more hysteresis is needed, the ±10-mV window can be made wider by increasing the reference voltages.

Sources

Chapter 1
Fig. 1-1. Radio-Electronics, 2/83, p. 76.
Fig. 1-2. Reprinted with the permission of National Semiconductor Corp., Linear Databook, 1982, p. 9-31.

Chapter 2
Fig. 2-1. Radio-Electronics, 6/89, p. 42.

Chapter 3
Fig. 3-1. Ham Radio, 11/78, p. 64.
Fig. 3-2. Reprinted with the permission of National Semiconductor Corp., Data Conversion/Acquisition Databook, 1980, p. 2-5.
Fig. 3-3. Signetics, Analog Data Manual, 1982, p. 16-28.
Fig. 3-4. Signetics, Analog Data Manual, 1982, p. 16-28.
Fig. 3-5. Signetics, 1987 Linear Data Manual Vol. 1: Communications, 11/86, p. 7-123.
Fig. 3-6. Radio-Electronics, 12/86, p. 57.
Fig. 3-7. Hands-On Electronics/Popular Electronics, 1/89, p. 97.

Fig. 3-8. Reprinted from EDN, 7/21/88, (c) 1989 Cahners Publishing Co., a division of Reed Publishing USA.
Fig. 3-9. Hands-On Electronics/Popular Electronics, 1/89, p. 84.
Fig. 3-10. Reprinted from EDN, 9/29/88, (c) 1989 Cahners Publishing Co., a division of Reed Publishing USA.
Fig. 3-11. Signetics, 1987 Linear Data Manual Vol. 1: Communications, 11/86, p. 7-123.
Fig. 3-12. Radio-Electronics, 3/86, p. 51.
Fig. 3-13. Signetics, 1987 Linear Data Manual Vol. 1: Communications, 11/86, p. 4-295.
Fig. 3-14. Signetics, Analog Data Manual, 1983, p. 11-10.
Fig. 3-15. EXAR, Telecommunications Databook, 1986, p. 9-23.

Chapter 4
Fig. 4-1. EXAR, Telecommunications Databook, 1986, p. 9-23.
Fig. 4-2. National Semiconductor Corp., Audio/Radio Handbook, 1980, p. 3-17.

Fig. 4-3. Signetics, 1987 Linear Data Manual, Vol. 1: Communications, 2/87, p. 4-66.

Fig. 4-4. Signetics, 1987 Linear Data Manual, Vol. 1: Communications, 11/86, p. 4-263.

Chapter 5

Fig. 5-1. Radio-Electronics, 12/86, p. 57.

Fig. 5-2. Radio-Electronics, 8/87, p. 63.

Fig. 5-3. Radio-Electronics, 8/87, p. 53.

Chapter 6

Fig. 6-1. Hands-On Electronics, 2/87, p. 92.

Chapter 7

Fig. 7-1. Linear Technology Corp., Linear Applications Handbook, 1987, p. AN5-5.

Fig. 7-2. Linear Technology Corp., Linear Applications Handbook, 1987, p. AN5-6.

Fig. 7-3. Linear Technology Corp., Linear Applications Handbook, 1986, p. 2-82.

Chapter 8

Fig. 8-1. Reprinted with permission of Control Engineering, 1301 S. Grove Ave., Barrington, Illinois, 12/73, p. 43.

Fig. 8-2. Courtesy of Motorola Inc., Communications Engineering Bulletin, EB-33.

Fig. 8-3. Courtesy, William Sheets.

Fig. 8-4. Hands-On Electronics, Sep/Oct 1986, p. 85.

Fig. 8-5. Courtesy, William Sheets.

Fig. 8-6. Linear Technology Design Notes, 3/89, Number 5.

Fig. 8-7. Linear Technology Corp., Linear Databook, 1986, p. 2-99.

Fig. 8-8. Reprinted from EDN, 8/75, (c) 1989 Cahners Publishing Co., a division of Reed Publishing USA.

Fig. 8-9. Reprinted from EDN, 9/73, (c) 1989 Cahners Publishing Co., a division of Reed Publishing USA.

Fig. 8-10. General Instrument Microelectronics, Application Note 1601, p. 3.

Fig. 8-11. Popular Electronics, Fact Card No. 65.

Fig. 8-12. Reprinted from EDN, 10/30/86, (c) 1989 Cahners Publishing Co., a division of Reed Publishing USA.

Fig. 8-13. GE/RCA, BiMOS Operational Amplifiers Circuit Ideas, 1987, p. 27.

Fig. 8-14. Courtesy of Motorola Inc., Communications Engineering Bulletin, EB-33.

Fig. 8-15. Supertex Data Book, 1983, p. 5-26.

Fig. 8-16. Reprinted with the permission of National Semiconductor Corp., Linear Applications Handbook, 1982, p. 9-75.

Fig. 8-17. Reprinted with the permission of National Semiconductor Corp., Linear Applications Handbook, 1982, p. 9-76.

Fig. 8-18. Reprinted from EDN, 10/30/86, (c) 1989 Cahners Publishing Co., a division of Reed Publishing USA.

Chapter 9

Fig. 9-1. Reprinted from EDN, 7/7/88, (c) 1989 Cahners Publishing Co., a division of Reed Publishing USA.

Fig. 9-2. Texas Instruments, Linear and Interface Circuits Applications, 1987, p. 12-9.

Fig. 9-3. Reprinted from EDN, 7/11/85, (c) 1989 Cahners Publishing Co., a division of Reed Publishing USA.

Fig. 9-4. Texas Instruments, Linear and Interface Circuits Applications, 1987, p. 12-8.

Fig. 9-5. Texas Instruments, Linear and Interface Circuits Applications, 1987, p. 12-10.

Fig. 9-6. Texas Instruments, Linear and Interface Circuits Applications, 1987, p. 12-8

Chapter 10

Fig. 10-1. Electronic Engineering, 12/75, p. 9.

Fig. 10-2. Radio-Electronics, 1979.

Fig. 10-3. Electronics Today International, 10/77, p. 47.

Chapter 11

Fig. 11-1. Radio-Electronics, 12/78, p. 77.

Fig. 11-2. Precision Monolithics Inc., 1981 Full Line Catalog, p. 8-12.

Fig. 11-3. Harris, Analog Product Data Book, 1988, p. 10-167.

Fig. 11-4. Siliconix, Integrated Circuits Data Book, 3/85, p. 5-16.

Chapter 12

Fig. 12-1. Courtesy, Williams Sheets.

Chapter 13

Fig. 13-1. Modern Electronics, 3/78, p. 68.

Fig. 13-2. Modern Electronics, 7/78, p. 55.

Fig. 13-3. Reprinted with permission from General Electric Semiconductor Department. General Electric SCR Manual, Sixth Edition, 1979, p. 226.

Fig. 13-4. Modern Electronics, 7/78, p. 55.

Fig. 13-5. Hands-On Electronics, Sep/Oct 1986, p. 24.

Fig. 13-6. Electronic Engineering, 9/86, p. 37.

Fig. 13-7. Reprinted from EDN, 2/21/85, (c) 1989 Cahners Publishing Co., a division of Reed Publishing USA.

Fig. 13-8. National Semiconductor Corp., Linear Applications Databook, p. 1079.

Fig. 13-9. TAB Books, The Build-It Book of Electronic Projects, p. 18.

Fig. 13-10. GE/RCA, BiMOS Operational Amplifiers Circuit Ideas, 1987, p. 23.

Fig. 13-11. NASA Tech Briefs, Spring 1983, p. 249.

Fig. 13-12. Courtesy of Texas Instruments Inc., Optoelectronics Databook, 1983-84, p. 15-9.

Fig. 13-13. Reprinted with the permission of National Semiconductor Corp., Linear Databook, 1982, p. 9-93.

Fig. 13-14. Electronic Design, 3/69, p. 96.

Fig. 13-15. Electronics Today International, 6/76, p. 43.

Fig. 13-16. Electronics Today International, 8/74, p. 66.

Fig. 13-17. Reprinted with the permission of National Semiconductor Corp., Linear Databook, 1982, p. 9-93.

Chapter 14

Fig. 14-1. Reprinted from Electronics, 12/77, p. 78. Copyright 1978 McGraw-Hill, Inc. All rights reserved.

Fig. 14-2. 101 Electronic Projects, 1977, p. 48.

Fig. 14-3. Courtesy, William Sheets.

Fig. 14-4. Courtesy, William Sheets.

Chapter 15

Fig. 15-1. Signetics, 1987 Linear Data Manual Vol. 2: Industrial, 2/87, p. 5-367.

Fig. 15-2. Intersil, Component Data Catalog, 1987, p. 5-112.

Fig. 15-3. Siliconix, Integrated Circuits Data Book, p. 85, p. 5-8.

Fig. 15-4. Signetics, RF Communications Handbook, 1989, p. 3-22.

Fig. 15-5. Reprinted from EDN, 1/20/79, (c) 1989 Cahners Publishing Co., a division of Reed Publishing USA.

Fig. 15-6. Courtesy of Motorola Inc. Linear Integrated Circuits, 1979, p. 6-98.

Fig. 15-7. (c) Siliconix Inc., Siliconix Application Note AN73-6, p. 4.

Fig. 15-8. Signetics, Analog Data Manual, 1983, p. 13-6.

Fig. 15-9. Reprinted with the permission of National Semiconductor Corp., National Semiconductor, Application Note LB-25.

Fig. 15-10. (c) Siliconix Inc., Siliconix Analog Switch & IC Product Data Book, 1/82, p. 6-9.

Chapter 16

Fig. 16-1. Radio-Electronics, 6/88, p. 49.

Fig. 16-2. Modern Electronics, 5/78, p. 6.

Fig. 16-3. R-E Experimenters Handbook, p. 162.

Fig. 16-4. Electronics Today International, 8/78, p. 61.

Fig. 16-5. Radio-Electronics, 2/84, p. 97.

Fig. 16-6. NASA, NASA Tech Briefs, p. 55.

Chapter 17

Fig. 17-1. Reprinted from EDN, 2/20/76, (c) 1989 Cahners Publishing Co., a division of Reed Publishing USA.

Chapter 18

Fig. 18-1. Signetics, Analog Data Manual, 1982, p. 3-76.

Fig. 18-2. Electronics Today International, 5/77, p. 77.

Fig. 18-3. Reprinted from Computers & Electronics. Copyright Ziff-Davis Publishing Company. 4/83, p. 109.

Fig. 18-4. Reprinted with permission from General Electric Semiconductor Department, General Electric SCR Manual, Sixth Edition, 1979, p. 440.

Fig. 18-5. Copyright by Computer Design. All rights reserved. Reprinted by permission. 1/83, p. 77.

Fig. 18-6. Reprinted with permission from General Electric Semiconductor Department, GE Semiconductor Data Handbook, Third Edition, p. 1371-4.

Fig. 18-7. Precision Monolithics Inc., Linear & Conversion IC Products, 7/78, p. 7-12.

Fig. 18-8. 101 Electronic Projects, 1977, p. 82.

Fig. 18-9. Reprinted with the permission of National Semiconductor Corp., Linear Databook, 1982, p. 3-109.

Fig. 18-10. Linear Technology Corp., Linear Databook Supplement, 1988, p. S2-34.

Fig. 18-11. Reprinted with the permission of National Semiconductor Corp., Data Conversion/Acquisition Databook, 1980, p. 3-88.

Fig. 18-12. Reprinted from EDN, 1/82, (c) 1989

Cahners Publishing Co., a division of Reed Publishing USA.

Chapter 19

Fig. 19-1. Electronics Today International, 6/79, p. 75

Fig. 19-2. Reprinted with the permission of National Semiconductor Corp., Linear Applications Handbook, 1982, p. LB33-1.

Fig. 19-3. Signetics, 555 Timers, 1973, p. 24.

Fig. 19-4. Reprinted with the permission of National Semiconductor Corp., Linear Databook, 1982, p. 9-143.

Chapter 20

Fig. 20-1. Hands-On Electronics, 3/87, p. 25.

Fig. 20-2. Motorola, TMOS Power FET Design Ideas, 1985, p. 17.

Fig. 20-3. Linear Technology Corp., Linear Applications Handbook, 1987, p. AN13-23.

Fig. 20-4. Intersil, Component Data Catalog, 1987, p. 5-113.

Fig. 20-5. Courtesy, William Sheets.

Fig. 20-6. Linear Technology Corp., Linear Databook, 1986, p. 3-23.

Fig. 20-7. Motorola Thyristor Device Data, Series A, 1985, p. 1-6-57.

Fig. 20-8. Motorola, Motorola TMOS Power FET Design Ideas, 1985, p. 16.

Chapter 21

Fig. 21-1. Reprinted from EDN, 3/21/85, (c) 1989 Cahners Publishing Co., a division of Reed Publishing USA.

Fig. 21-2. Electronic Engineering, 11/86, p. 39.

Fig. 21-3. General Electric, Application Note 90.16, p. 27.

Fig. 21-4. Precision Monolithics Inc., 1981 Full Line Catalog, p. 14-17.

Fig. 21-5. Precision Monolithics Inc., 1981 Full Line Catalog, p. 14-17.

Fig. 21-6. GE/RCA, BiMOS Operational Amplifiers Circuit Ideas, 1987, p. 12.

Fig. 21-7. National Semiconductor Corp., Linear Databook, 1982, p. 3-97.

Fig. 21-8. Radio-Electronics, 4/89, p. 60.

Fig. 21-9. Reprinted from EDN, 12/8/88, (c) 1989 Cahners Publishing Co., a division of Reed Publishing USA.

Fig. 21-10. Texas Instruments, Linear and Interface Circuits Applications, Vol. 1, 1985, p. 3-18.

Fig. 21-11. Reprinted with the permission of National Semiconductor Corp., Linear Databook, 1982, p. 3-97.

Fig. 21-12. RCA, Solid State Division, Digital Integrated Circuits Application Note ICAN-6346, p. 5.

Fig. 21-13. General Electric/RCA, BiMOS Operational Amplifiers Circuit Ideas, 1987, p. 18.

Fig. 21-14. Reprinted with permission of Analog Devices, Inc. Data Acquisition Databook, 1982, p. 4-123.

Fig. 21-15. Reprinted from EDN, 10/17/85, (c) 1989 Cahners Publishing Co., a division of Reed Publishing USA.

Fig. 21-16. Teledyne Semiconductor, Data Acquisition IC Handbook, 1985, p. 15-15.

Fig. 21-17. Courtesy of Fairchild Camera & Instrument Corporation. Linear Databook, 1982, p. 5-38.

Fig. 21-18. Electronics Today International, 3/78, p. 50.

Fig. 21-19. Popular Electronics, 3/79, p. 78.

Fig. 21-20. Courtesy of Fairchild Camera & Instrument Corporation. Linear Databook, 1982, p. 5-38.

Chapter 22

Fig. 22-1. Electronic Engineering, 2/86, p. 38.

Fig. 22-2. Courtesy, William Sheets.

Fig. 22-3. Electronic Engineering, 4/77, p. 13.

Fig. 22-4. Electronic Engineering, 7/85, p. 34.

Fig. 22-5. Electronic Design, 3/77, p. 106.

Fig. 22-6. Electric Engineering, 1/87, p. 25.

Chapter 23

Fig. 23-1. Modern Electronics, 3/78, p. 50.

Fig. 23-2. 73 Amateur Radio, 6/83, p. 106.

Fig. 23-3. TAB Books, 104 Weekend Electronics Projects, p. 70.

Fig. 23-4. Courtesy, William Sheets.

Fig. 23-5. 73 Amateur Radio, 9/75, p. 105.

Chapter 24

Fig. 24-1. Hands-On Electronics, 1-2/86, p. 96.

Fig. 24-2. Popular Electronics/Hands-On Electronics, 5/89, p. 85.

Fig. 24-3. Hands-On Electronics, Fact Card No. 57.

Fig. 24-4. RF Design, 12/86, p. 41.

Fig. 24-5. Hands-On Electronics, 7/87, p. 47.

Fig. 24-6. Radio-Electronics, 1/67.

Fig. 24-7. (c) Siliconix Inc., T100/T300 Applications.

Fig. 24-8. Reprinted with permission from General

Electric Semiconductor Department, General Electric SCR Manual, Sixth Edition, 1979, p. 224.

Fig. 24-9. Popular Electronics, 11/77, p. 62.

Fig. 24-10. Electronics Today International, 6/82, p. 69.

Fig. 24-11. Siliconix, Small-Signal FET Data Book, 1/86, p. 7-28.

Chapter 25

Fig. 25-1. Reprinted from EDN, 4/76, (c) 1989 Cahners Publishing Co., a division of Reed Publishing USA.

Fig. 25-2. General Electric, Application Note 90.16, p. 26.

Fig. 25-3. Courtesy of Fairchild Camera & Instrument Corporation. Linear Databook, 1982, p. 5-25.

Fig. 25-4. Radio-Electronics, 3/89, p. 12.

Fig. 25-5. Texas Instruments, Linear and Interface Circuits Applications, 1985, Vol. 1, p. 7-11 and 7-12.

Fig. 25-6. Signetics, 555 Timers, 1973, p. 17.

Fig. 25-7. Reprinted from EDN, 9/19/85, (c) 1989 Cahners Publishing Co., a division of Reed Publishing USA.

Fig. 25-8. GE, Application Note 90.16, p. 26.

Chapter 26

Fig. 26-1. Electronic Engineering, 12/75, p. 15.

Fig. 26-2. Radio-Electronics, Experimenters Handbook, p. 122.

Chapter 27

Fig. 27-1. Electronic Engineering, 5/76, p. 17.

Fig. 27-2. Reprinted from Electronics, 8/78, p. 106. Copyright 1978, McGraw-Hill Inc. All rights reserved.

Fig. 27-3. Electronic Design, 4/74, p. 114.

Fig. 27-4. Electronic Engineering, 10/86, p. 41.

Fig. 27-5. Reprinted with the permission of National Semiconductor Corp., Hybrid Products Databook, 1982, p. 2-15.

Fig. 27-6. 49 Easy to Build Projects, TAB Book No. 1337, p. 77.

Fig. 27-7. Electronics Today International, 1/79, p. 97.

Fig. 27-8. Reprinted with the permission of National Semiconductor Corp., Hybrid Products Databook, 1982, p. 2-16.

Chapter 28

Fig. 28-1. Courtesy, William Sheets.

Fig. 28-2. Electronics Today International, 12/78, p. 93.

Fig. 28-3. Radio-Electronics, 4/87, p. 48.

Fig. 28-4. Electronics Today International, 4/78, p. 63.

Chapter 29

Fig. 29-1. TAB Books, 303 Dynamic Electronic Circuits, p. 169.

Fig. 29-2. Electronics Today International, 12/77, p. 86.

Chapter 30

Fig. 30-1. Ham Radio, 1/84, p. 94.

Fig. 30-2. Reprinted with permission from General Electric Semiconductor Department, Optoelectronics, Second Edition, p. 119.

Fig. 30-3. Signetics, Analog Data Manual, 1982, p. 16-27.

Fig. 30-4. Modern Electronics, 7/78, p. 56.

Fig. 30-5. Electronic Engineering, 2/87, p. 40.

Fig. 30-6. GE, Optoelectronics, Third Edition, Ch. 6, p. 124.

Fig. 30-7. 73 Amateur Radio, 1/84, p. 115.

Fig. 30-8. Electronic Design, 12/78, p. 95.

Chapter 31

Fig. 31-1. NASA, Tech Briefs, Spring 1985, p. 40.

Fig. 31-2. NASA, Tech Briefs, Fall/Winter 1981, p. 319.

Fig. 31-3. Popular Electronics, 12/82, p. 82.

Chapter 32

Fig. 32-1. Hands-On Electronics, 9/87, p. 88

Fig. 32-2. Texas Instruments, Linear and Interface Circuits Applications, Vol. 1, 1985, p. 7-15.

Fig. 32-3. TAB Books, The Giant Book of Easy-to-Build Electronics Projects, 1982, p. 31.

Fig. 32-4. Popular Electronics, 9/89, p. 88.

Fig. 32-5. Popular Electronics, 9/89, p. 88.

Fig. 32-6. Hands-On Electronics, 9/87, p. 89.

Fig. 32-7. Hands-On Electronics, 9/87, p. 89.

Fig. 32-8. National Semiconductor Corp., Transistor Databook, 1982, p. 7-11.

Fig. 32-9. Popular Electronics, 7/89, p. 25.

Fig. 32-10. Courtesy, William Sheets.

Fig. 32-11. Courtesy, William Sheets.

Fig. 32-12. Hands-On Electronics, Fact Card No. 86.

Fig. 32-13. (c) Siliconix Inc., Siliconix Application Note AN154.

Fig. 32-14. Wireless World, 5/78, p. 69.

Fig. 32-15. Modern Electronics, 2/78, p. 16.

Fig. 32-16. Reprinted with permission from General

Electric Semiconductor Department, 2/68.

Fig. 32-17. Electronics Today International, 6/74, p. 67.

Fig. 32-18. Electronics Today International, 4/78, p. 81.

Fig. 32-19. Modern Electronics, 6/78, p. 14.

Fig. 32-20. Modern Electronics, 2/78, p. 17.

Fig. 32-21. Electronics Today International, 5/75, p. 68.

Fig. 32-21. Reprinted with permission from General Electric Semiconductor Department, 2/68.

Fig. 32-22. Reprinted with the permission of National Semiconductor Corp., Linear Databook, 1982, p. 9-143.

Chapter 33

Fig. 33-1. Reprinted with the permission of National Semiconductor Corp., Linear Databook, 1982, p. 3-204.

Fig. 33-2. 73 Amateur Radio, 10/77, p. 115.

Fig. 33-3. 104 Weekend Electronics Projects, TAB Book No. 1436, p. 64.

Fig. 33-4. Electronics Today International, 1975, p. 72.

Fig. 33-5. Electronics Today International, 17/81, p. 75.

Fig. 33-6. Reprinted with permission from General Electric Semiconductor Department, GE Application Note 200.35, 3/66, p. 14.

Fig. 33-7. Courtesy, William Sheets.

Fig. 33-8. Radio-Electronics, 12/83, p. 38.

Chapter 34

Fig. 34-1. Electronics Today International, 6/76, p. 40.

Fig. 34-2. Signetics, Analog Data Manual, 1982, p. 8-10.

Fig. 34-3. Reprinted from EDN, 5/73, (c) 1989 Cahners Publishing Co., a division of Reed Publishing USA.

Fig. 34-4. Fairchild Camera and Instrument Corp., Linear Databook, 1982, p. 4-180.

Fig. 34-5. Courtesy of Motorola Inc., Linear Integrated Circuits, 1979, p. 6-123.

Fig. 34-6. Texas Instruments, Linear and Interface Circuits Applications, 1985, Vol. 1, p. 3-46 and 3-47.

Fig. 34-7. Electronic Engineering, 1/83, p. 31.

Fig. 34-8. Reprinted from EDN, 8/20/78, (c) 1989 Cahners Publishing Co., a division of Reed Publishing USA.

Fig. 34-9. Electronic Engineering, 7/88, p. 27.

Fig. 34-10. Signetics, Analog Data Manual, 1977, p. 264.

Chapter 35

Fig. 35-1. Electronics Today International, 8/78, p. 69.

Fig. 35-2. Courtesy of Fairchild Camera and Instrument Corporation, Linear Databook, 1982, p. 5-32.

Fig. 35-3. (c)Siliconix Inc., Analog Switch & IC Product Data Book, 1/82, p. 6-18.

Fig. 35-4. Courtesy of Motorola Inc., Linear Integrated Circuits, 1979, p. 6-123.

Fig. 35-5. Courtesy of Texas Instruments Inc., Linear Control Circuits Data Book, Second Edition, p. 205.

Fig. 35-6. (c)Siliconix Inc., Analog Switch & IC Product Data Book, 1/82, p. 6-14.

Fig. 35-7. Texas Instruments, Linear and Interface Circuits Applications, Vol. 1, 1985, p. 3-23.

Index